Explosion, Shock-Wave and HIGH-STRAIN-RATE PHENOMENA OF ADVANCED MATERIALS

SERIES MULTIPHYSICS: ADVANCES AND APPLICATIONS

Explosion, Shock-Wave and HIGH-STRAIN-RATE PHENOMENA OF ADVANCED MATERIALS

Edited by

KAZUYUKI HOKAMOTO

Professor, Institute of Industrial Nanomaterials, Kumamoto University, Kumamoto, Japan

ELSEVIER

ACADEMIC PRESS

An imprint of Elsevier

elsevier.com/books-and-journals

Academic Press is an imprint of Elsevier
125 London Wall, London EC2Y 5AS, United Kingdom
525 B Street, Suite 1650, San Diego, CA 92101, United States
50 Hampshire Street, 5th Floor, Cambridge, MA 02139, United States
The Boulevard, Langford Lane, Kidlington, Oxford OX5 1GB, United Kingdom

Notices

Knowledge and best practice in this field are constantly changing. As new research and experience broaden our understanding, changes in research methods, professional practices, or medical treatment may become necessary.

Practitioners and researchers must always rely on their own experience and knowledge in evaluating and using any information, methods, compounds, or experiments described herein. In using such information or methods they should be mindful of their own safety and the safety of others, including parties for whom they have a professional responsibility.

To the fullest extent of the law, neither the Publisher nor the authors, contributors, or editors, assume any liability for any injury and/or damage to persons or property as a matter of products liability, negligence or otherwise, or from any use or operation of any methods, products, instructions, or ideas contained in the material herein.

British Library Cataloguing-in-Publication Data

A catalogue record for this book is available from the British Library

Library of Congress Cataloging-in-Publication Data

A catalog record for this book is available from the Library of Congress

ISBN: 978-0-12-821665-1

For Information on all Academic Press publications visit our website at https://www.elsevier.com/books-and-journals

Publisher: Matthew Deans
Acquisitions Editor: Carrie Bolger
Editorial Project Manager: Charlotte Rowley
Production Project Manager: Kamesh Ramajogi
Cover Designer: Alan Studholme

Typeset by Aptara, New Delhi, India

Contents

Contributors

Ivan Bataev
Novosibirsk State Technical University, Russia

Hideki Hamashima
Kumamoto Industrial Research Institute, Japan

Kazuyuki Hokamoto
Professor, Institute of Industrial Nanomaterials, Kumamoto University, Kumamoto, Japan

Hirofumi Iyama
National Institute of Technology, Kumamoto College, Kumamoto, Japan

Akihisa Mori
Sojo University, Kumamoto, Japan

Masatoshi Nishi
National Institute of Technology, Kumamoto College, Kumamoto, Japan

Nejc Novak
Faculty of Mechanical Engineering, University of Maribor, Maribor, Slovenia

Zoran Ren
Faculty of Mechanical Engineering, University of Maribor, Maribor, Slovenia; International Research Organization for Advanced Science and Technology, Kumamoto University, Kumamoto, Japan

Shigeru Tanaka
Kumamoto University, Japan

Matej Vesenjak
Faculty of Mechanical Engineering, University of Maribor, Maribor, Slovenia

Naoyuki Wada
Asahi Kasei Corporation, Chikushino, Japan

About the Authors

Hideki Hamashima has been working at Kumamoto Industrial Research Institute as a Researcher since 2008. He received a Ph.D. degree from Kumamoto University in 2004 on "Study on the Detonation Behavior of Ideal and Non-Ideal Explosives" followed by his postdoc experiences at Kumamoto University and Japan's National Institute of Advanced Industrial Science & Technology, Tsukuba. He has been engaged in the numerical simulation and measurement of various explosion and shock-wave phenomena.

Hirofumi Iyama found his position at Yatsushiro National College of Technology in 1997 and received a Ph.D. degree from Kumamoto University in 2002. The College was renamed as National Institute of Technology, Kumamoto College in 2009. He has been conducting research on explosive forming, high-velocity deformation of metal plates, and other processing technologies using underwater shockwaves. He has also worked as a Director of the Institute of Shockwave Advanced Technology since 2017 and strives to develop and contribute to society through the application of shockwave technology.

Masatoshi Nishi is a Lecturer at the National Institute of Technology, Kumamoto College. He graduated from Kumamoto University in the spring of 2011 and received a Ph.D. degree in Engineering followed by Postdoctoral research for 1 year at the Institute of Pulsed Power Science, Kumamoto University. His research interests include explosives and other impulsive metal forming and welding, and he is now engaged in the computational analysis of high-strain-rate deformation.

Shigeru Tanaka has been employed at Kumamoto University, originally as a technical staff member, since 2004, and he has been involved in the optical observation of various explosion phenomena at the explosion facility. He received a Ph.D. degree in engineering from Kumamoto University in 2009 and is currently working as Assistant Professor at the Institute of Industrial Nanomaterials, Kumamoto University. He has authored more than 60 journal papers, mostly on explosive materials processing. He is now interested in developing a process using explosive-driven underwater shockwaves for micro- or sub-micrometer order plate forming and punching.

Akihisa Mori graduated from Kumamoto University in 1998, received a Master's degree from Sojo University in 2003, and doctoral degree in engineering from Kumamoto University, Japan, in 2006. He has been active in the fields of the explosion, shockwaves, and high-strain-rate phenomena at the Research Center for Advances in Impact Engineering, Sojo University since 2009 and is now an Associate Professor in the Department of Mechanical Engineering, Faculty of Engineering, Sojo University. He received the Excellent Presentation Promotion Award from the Japan Society for Technology of Plasticity in 2005 and 2010.

Ivan Bataev received a Ph.D. degree from Novosibirsk State Technical University (NSTU) in 2010. In 2019, he received a D.Sc. degree. His research interests include understanding the structural evolution during extreme external impacts on materials. He mainly concentrates on a metallographic analysis of explosively welded materials as well as on numerical simulation of the high-velocity collision of materials. He has published more than 50 journal papers and holds several patents. Currently, he works as a Professor in the Materials Science and Engineering Department of NSTU and heads the research laboratory "Physicochemical technologies and functional properties of materials."

Naoyuki Wada received a Ph.D. degree from Kumamoto University in 2013 on the synthesis of various ceramics and intermetallics through under-liquid explosion phenomena. He is currently working as Assistant Manager of Quality Assurance-Chikushino Group, Safety Environment & Quality Assurance Department, Explosives Division, Asahi Kasei Corporation, and he has been engaged in R&D and quality assurance of explosion bonded composites since 2013.

Zoran Ren joined the University of Maribor, Slovenia, in 1987, after receiving the Bachelor degree in Mechanical Engineering from the same university. He moved to Swansea University College, United Kingdom, in 1990 and received a Ph.D. degree in Advanced Computational Engineering from the University of Wales, United Kingdom, in 1994. He returned to the University of Maribor, Slovenia, in 1993, to pursue his academic career and was appointed Full Professor of Mechanical Engineering in 2005. He has also been an Alexander von Humboldt Research Fellow since 2000 and a Visiting Professor at Kumamoto University, Japan, since 2016. His research interests include studying the behavior of engineering materials under high deformation rates through analytical, computational, and experimental methods. He has authored more than 1000 bibliographical documents comprising journal papers, conference contributions, books, studies, research findings, patents, etc. His numerous joint scientific publications reflect a consistent record of good international cooperation. He is currently Vice-Rector of Science and Research at the University of Maribor.

Matej Vesenjak received a Ph.D. degree from the Faculty of Mechanical Engineering, University of Maribor, Slovenia, in 2006, where he has been a Full Professor since 2018. His research interests include the design and characterization of advanced cellular structures and metamaterials with a strong emphasis on geometrical properties, experimental testing, and computational simulations. He has published more than 120 original scientific journal articles and has contributed to more than 250 conference papers. He has a strong record of international cooperation and has given more than 70 invited lectures at universities around the world. He is also a Visiting Professor and has received many national and international fellowships and awards for his academic achievements.

Nejc Novak is an Assistant Professor in the Faculty of Mechanical Engineering, University of Maribor, Slovenia. His research interests include advanced computational modeling and simulations, cellular structures, auxetic cellular materials, characterization and optimization of cellular structures, multi-physical problems, crashworthiness, and blast and ballistic loading. He was promoted to Doctor of Philosophy in 2019 in the University of Maribor, Slovenia, with the thesis title "Computational simulations of auxetic cellular structures." His bibliography comprises more than 85 bibliographical units, of which more than 18 original scientific papers have been published mostly in high-ranked journals in the field of composites and mechanics. He has been a Visiting Researcher at the Kumamoto University, Japan, at the Georgia Institute of Technology, USA, at the Beijing Institute of Technology, China, as well as at the Politecnico di Milano, Italy.

Foreword

The use of explosives and other intense dynamic loading is normally considered difficult especially for applying materials processing technology due to the high-rate deformation over a short period. It is known that the interaction of shockwaves may induce problems for the recovery of materials. I have engaged on related topics, and I believe that it is possible to know the phenomena accurately so as to improve their recovery. Based on the measurements and simulations being rapidly developed, lots of new technologies are expected to be industrialized. As an example, I have tried to use underwater shockwaves for food processing using electrical wire explosion. Though the energetic source is different, the basics are almost the same. The knowledge based on the experiences may bring new ideas to improve the processes. I hope this book is informative on the unique technology for materials processing and the responses of materials under high-rate deformation.

Shigeru Itoh
Professor Emeritus,
Kumamoto University, Kumamoto, Japan
Representative Director, Institute of Shockwave
Advanced Technology, Kumamoto, Japan
Formerly President of National Institute of Technology,
Okinawa College, Okinawa, Japan

Preface

It is my great pleasure to issue this book named "Explosion, Shock-wave and High-strain-rate Phenomena of Advanced Materials" based on the international collaborations made by the contributing authors.

Kumamoto University has a research facility to conduct experiments using explosives, and the intense dynamic loading released by explosions has been used for various materials processing technologies since the 1970s. The Shock Wave and Condensed Matter Research Center was opened in 1999, and an explosion facility with two explosion pits, currently used for various measurement and recovery experiments, was constructed in 2001. The facility is now under the Institute of Industrial Nanomaterials, is unique among the national universities of Japan, and is open for use by researchers.

The authors of each chapter have conducted collaborative research works using the explosion facility over a long period, and the chapters are written based on the collaborative works and the achievements of each author.

The book includes basic issues and some recent research results on explosive forming, welding, and the making of unique porous materials and their responses. I hope the book will be useful for researchers and engineers in related fields.

Kazuyuki Hokamoto
Professor
Institute of Industrial Nanomaterials
Kumamoto University
Kumamoto, Japan

CHAPTER 1

Basic issues in explosion and other high-rate processing of materials

Kazuyuki Hokamoto[a], Hideki Hamashima[b]
[a]Professor, Institute of Industrial Nanomaterials, Kumamoto University, Kumamoto, Japan
[b]Kumamoto Industrial Research Institute, Japan

1.1 Introduction

This chapter presents basic issues of explosion phenomena and demonstrates some applications of high-rate materials processing technology. It is important to know the basic issues of such explosion phenomena, such as keywords, important phenomena, and scaling law in both air and water. Since the phenomena are dynamic or high-rate processes, which are different from static phenomena, some basic issues are explored in the first part of this chapter.

After discussion of the basic issues, some applications for high-rate materials processing technology are briefly introduced. Take explosive welding as an example: it is normally performed under direct contact of an explosive layer over a metal flyer plate because the detonation pressure is not transferred effectively when there is a space or gap between them. In addition, we sometimes use water as a pressure transmitting medium because it is possible to transmit high pressure as an underwater shock wave to the material to be processed. The use of water as a pressure transmitting medium is technically important for controlling the pressuring condition of the materials to be accelerated. Based on the change of the assembly in dimensions, many pressurizing devices employed for such high-rate processing have been developed and investigated using the explosion facility at Kumamoto University. This book demonstrates numerous case studies based on theoretical and numerical analyses, including the rate effects.

Explosion, Shock-Wave and High-Strain-Rate Phenomena of Advanced Materials
DOI: 10.1016/C2019-0-03954-4

1.2 Explosion phenomena

1.2.1 Explosion in air

1.2.1.1 Blast

When an explosive explodes in air, high-pressure and high-temperature gas is generated by the explosion, which compresses the surrounding air. It spreads out at a very high speed and creates a shock wave. The shock wave spreads, making a spherical surface around the point of initiation, and compresses and flows as a thin layer in the air, which can destroy materials and structures. The pressure rises sharply at the shock front. The fluctuation state of the air caused by the explosion phenomenon of the explosive is called the *blast* or *blast wave*.

1.2.1.2 Structure and propagation of a blast wave

Fig. 1.1 shows a typical pressure–time profile (waveform) of a blast arriving at a specific point at a certain distance from the explosion point. The peak over pressure is decreased when the distance from the explosion point is increased. The peak over pressure and the impulse of positive phase are the important parameters for discussing the blasting effect. The important elements of the blast are explained as follows:

1. *Shock front*: The front of the shock wave generated by an explosion is called the *shock front*, which is a very thin wave front. It involves a very sharp rise in pressure. The density, temperature, and pressure of the air before and after the shock front are discontinuous.
2. *Peak over pressure*: The maximum pressure from the steep rise induced from the shock wave is called *peak over pressure* and is above normal atmospheric pressure.

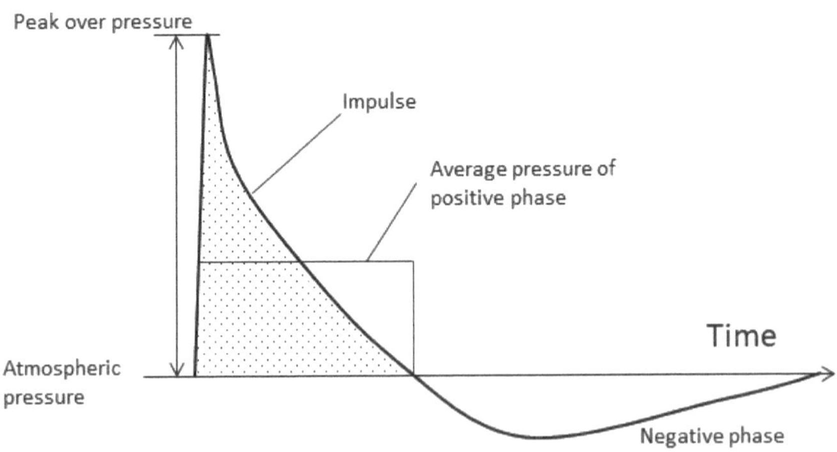

Fig. 1.1 ***Typical pressure–time profile (waveform) of a blast wave.***

3. *Positive or pressure phase*: At the shock front, the pressure rises sharply from atmospheric pressure to peak over pressure but falls with time and returns to atmospheric pressure again. The part of this pressure higher than the atmospheric pressure is called the *positive phase* or *pressure phase*.
4. *Negative or suction phase*: Air and explosion gas flow outward as a strong wind behind the shock wave. The strong wind creates a negative pressure below atmospheric pressure. Due to the inertia of the wind, the negative pressure phase spreads outward after the positive phase. Therefore, observed at a specific point, after the positive phase has passed, the direction of the wind changes and it blows toward the initiation point. The maximum negative pressure is a fraction of the peak over pressure. The portion below atmospheric pressure following the positive phase is called the *negative phase* or *suction phase*.
5. *Particle velocity*: As the shock wave travels through the air, not only do air pressure, air density, and temperature behind the wave front increase but also air particles (molecules) accelerate, giving them particle velocity (wind velocity).

1.2.1.3 Reflection of blast wave

When the shock wave collides vertically with a rigid wall, it is reflected by the wall. After the collision, the pressure, density, and temperature behind the shock wave are greater than those of the incident wave. Although the reflected pressure is theoretically 2 to 8 times higher than the incident wave, some literature suggests the pressure is more than 20 times, as it is not a perfect gas [1].

Fig. 1.2 shows the state where a plane shock wave collides with a rigid wall at a certain angle and is reflected. When the incident shock wave I collides with the wall at the incident angle α_I and is reflected at the reflection angle α_R (α_I and α_R are not necessarily equal), the state changes from the state of (A) to the state of (C) on the wall.

Fig. 1.3 shows the situation when the reflected wave from the ground surface catches up with the first shock wave under a strong shock wave, such as with the detonation of an explosive. At that point, a third shock

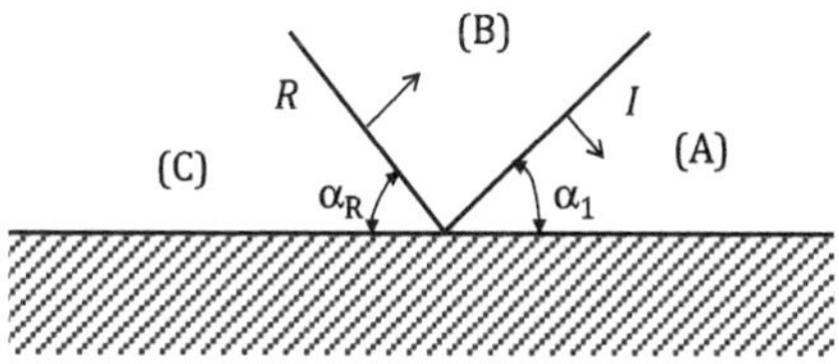

Fig. 1.2 ***Oblique reflection of a plane shock wave.***

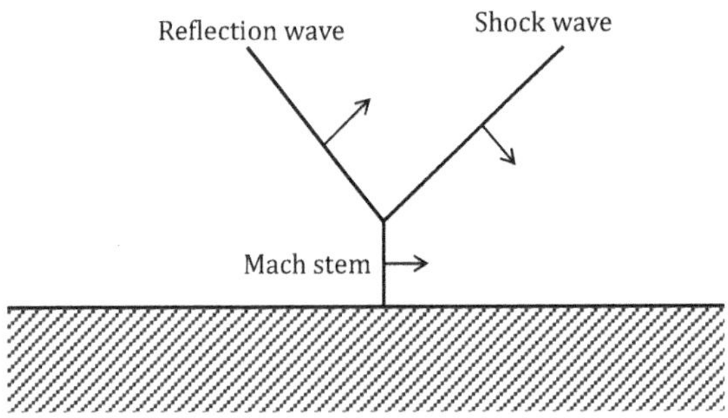

Fig. 1.3 ***Reflection of a strong shock wave inducing Mach stem.***

wave with a wave front vertical to the ground surface is formed. This third shock wave is called a *Mach stem*. The Mach stem increases its height as it moves sidewise. At the triple point, the peak over pressure and the impulse are both at a maximum. The incident pressure and the reflected pressure at the same distance from the ground are absorbed and integrated, respectively. This Mach stem phenomenon is called *Mach reflection*. If the explosion height is selected properly, the effect of Mach reflection can be increased by more than 50% compared with exploding the same explosive on the ground [2]. However, weak shock waves near the sound velocity are reflected symmetrically on the ground surface, and the Mach reflection phenomenon is not induced.

1.2.1.4 Scaling law of blast pressure and impulse

Blast pressure propagating in air has been discussed based on the scaling law, and it is well known that the peak over pressure and the impulse depend on the scaling distance Z ($m{\cdot}kg^{-1/3}$), where $Z = R/M^{1/3}$ (M: mass of the explosive, R: distance from the center of explosive). Early works by Kinney and Graham [2], Kingery and Pannill [3], and Baker [4] are well known, and a recent work by Arai and Hiyoshi [5] on measurement experiments for TNT and emulsion explosives are illustrated in Figs. 1.4 and 1.5. The line in the figures, NRIPS09, is the fitted data curve.

1.2.2 Underwater explosion

The basic phenomena of underwater explosion can be divided in two. One is shock waves propagating in water, and the other is the explosion product causing dynamic motion of gas bubbles [6]. If the former shock wave does not reflect on the water surface or bottom, it will travel far away without returning. In the latter, product gas repeatedly expands and contracts, and exhibits complicated behavior such that a new shock wave

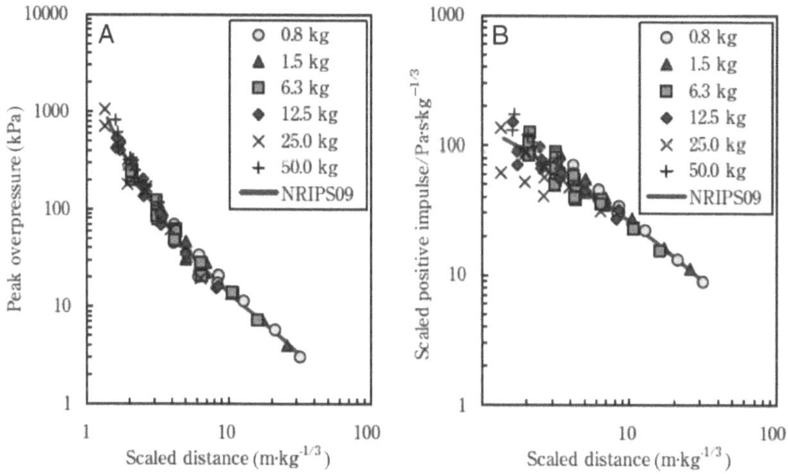

Fig. 1.4 ***Peak over pressure (A) and impulse (B) of a blast wave for a TNT explosive.*** *(From H. Arai and R. Hiyoshi. 2011. Blast pressure measurement of TNT and commercial explosives above the ground. [In Japanese.]* Science and Technology of Energetic Materials *72(1) 1–8).*

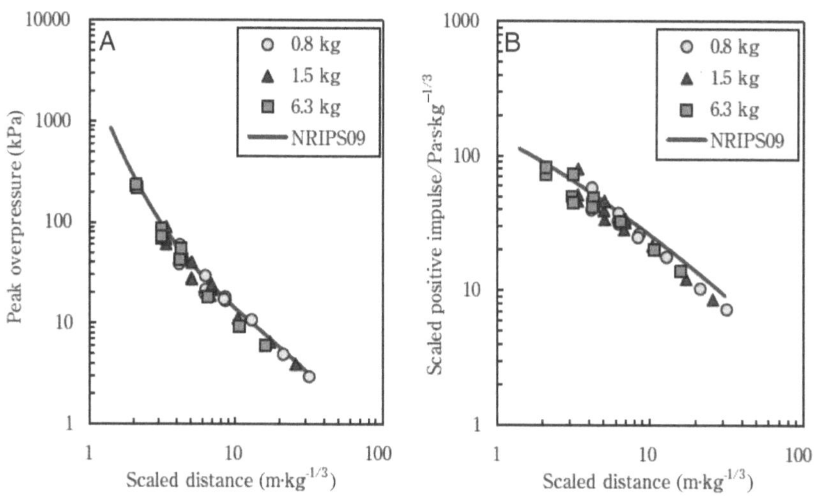

Fig. 1.5 ***Peak over pressure (A) and impulse (B) of a blast wave for an emulsion explosive.*** *(From H. Arai and R. Hiyoshi. 2011. Blast pressure measurement of TNT and commercial explosives above the ground. [In Japanese.]* Science and Technology of Energetic Materials *72(1) 1–8).*

is generated at the time of contraction. When exploding in a small water container, the explosion product gas is released easily into the air. In this case, it is not necessary to consider the effect of the bubble and is possible to think only of the effect of the generated shock wave. However, when the explosion product gas is not released into the air, such as when exploded in deep water, it is necessary to consider the influence of the behavior of the explosion product gas called the *bubble pulse*. It is also said that the bubble pulse may affect the explosive forming process using underwater shock waves [7].

1.2.2.1 Peak over pressure and impulse of underwater shock wave

The peak over pressure in underwater explosion is much higher than in the air, as the higher density can transfer the shock wave more efficiently. For this reason, underwater shock waves are sometimes used for materials processing, as mentioned in the following part of this book. According to the data of AMCP706-180 [8], the shock waves are 234, 15.2, and 1.1 MPa at distances of 1.5, 15, and 152 m, respectively, when 136 kg of TNT is exploded in water. During underwater explosion, the explosive gas expands and contracts repeatedly. The peak over pressure and the impulse of the underwater shock wave are given by Eq. (1.1) and Eq. (1.2) [6], the same as for an explosion in air:

$$P_m = K\left(M^{1/3} / R\right)^{\alpha} \tag{1.1}$$

$$I_m = LM^{1/3}\left(M^{1/3} / R\right)^{\beta} \tag{1.2}$$

where P_m is the peak pressure (MPa); I_m is the impulse (ms·MPa); M is the mass of the explosive (kg); R is the distance from the center of the explosive (m); and K, L, α, and β represent the characteristic value of the explosive for the underwater shock wave [6].

1.2.2.2 Behavior of the bubble pulse in water

Fig. 1.6 shows a schematic diagram of the underwater explosion phenomenon. The first intense shock wave is induced by the explosion and spreads immediately out at high velocity. The expanding gas spreads relatively slowly, in the order of milliseconds, to make a spherical bubble. The direction of expanding gas changes to cause shrinkage of the gas when the outside water pressure becomes higher than the gas phase. Due to the shrinkage, the gas contracts to a point or a small volume that causes a secondary shock wave to spread out again. The whole procedure is called the *bubble pulse phenomena*.

It is possible to visualize the phenomena using a high-speed camera and through numerical simulation. Fig. 1.7 shows high-speed photography images when using a 2-g SEP explosive, a PETN-based explosive ignited

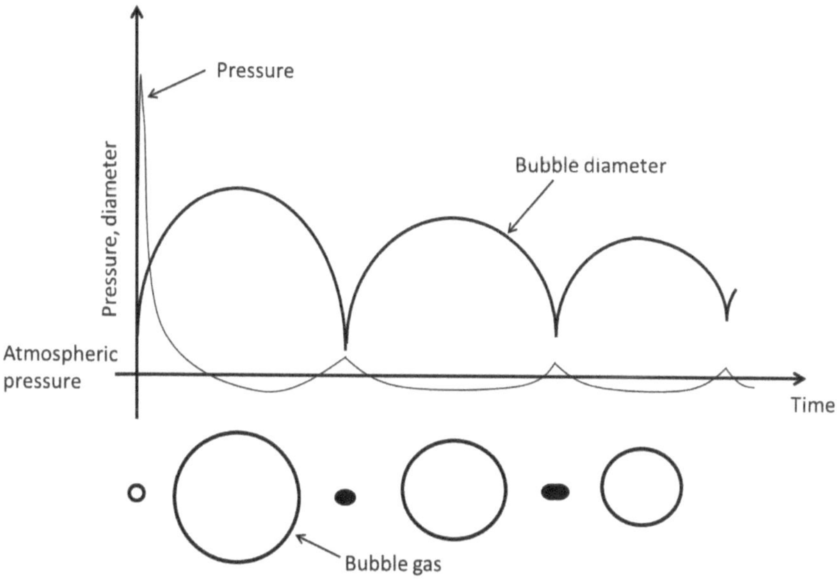

Fig. 1.6 ***Schematic illustration of underwater explosion phenomena.***

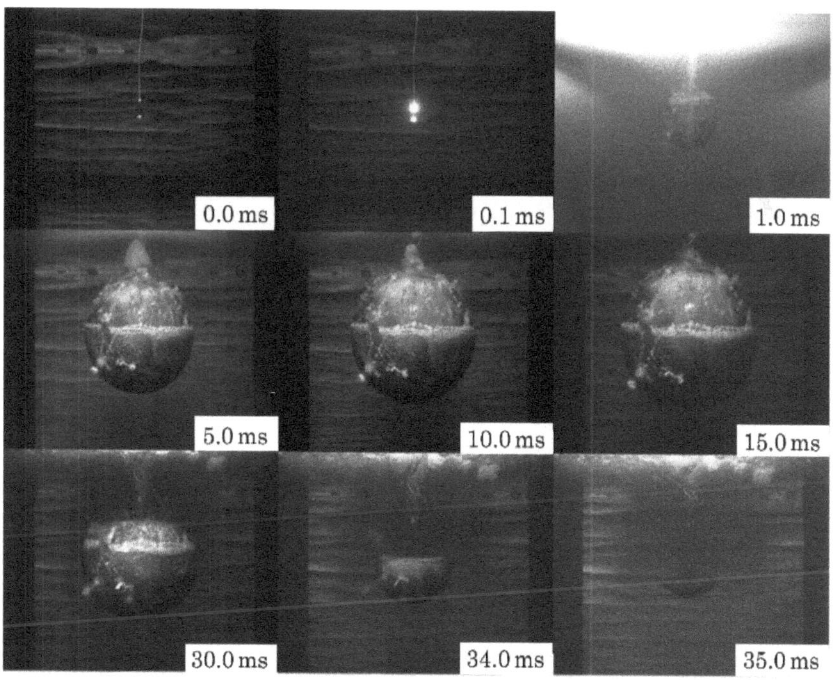

Fig. 1.7 ***Visualization of bubble pulse motion using a high-speed camera.*** *(From K. Hokamoto. 2018. Challenges for unexplored materials processing technology. [In Japanese.]* Bulletin of the Japan Society for Technology of Plasticity *1(4): 235–236).*

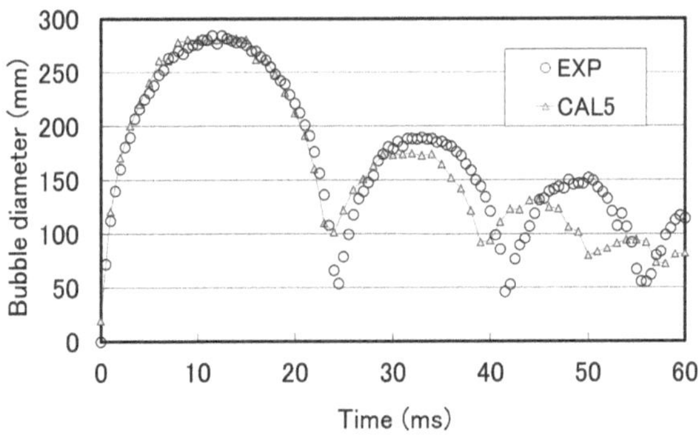

Fig. 1.8 ***Numerically simulated result of the size of a bubble with time.*** *(From H. Hamashima, M. Shibuta, and S. Itoh. 2011. Behavior of bubble in small water tank used for food processing using underwater shock wave.* Materials Science Forum *673 225–230).*

by an electric detonator [9]. Fig. 1.8 shows the numerically simulated result of the size of the bubble with time for a case using a 0.6-g SEP explosive [10]. The Jones–Wilkins–Lee equation for the explosive is cited in the work of Itoh et al. [11]. The calculation was made using LS-DYNA code and shows good agreement with the measurement results [10]. Fig. 1.9 shows the position (depth) of the bubble and that the bubble moved downward [10]. It should be noted that the motion is due to a flow toward the wall at the bottom that is close to the detonation point.

The bubble pulse has a feature that the maximum peak pressure is decreased after the initial shock wave, but the pressurization time becomes longer. The secondary shock wave effect is considered important for

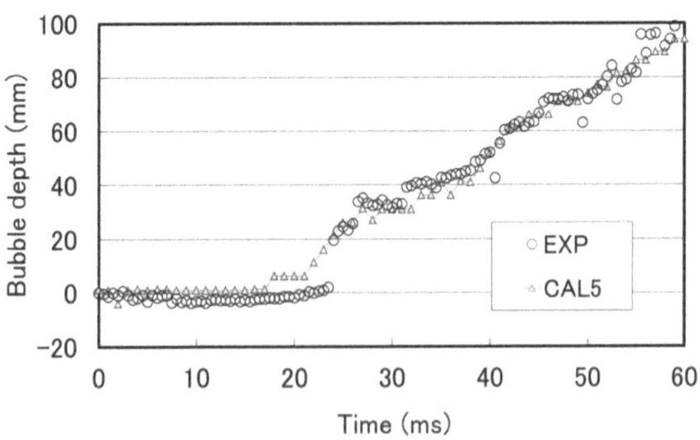

Fig. 1.9 ***Position of the bubble moving downward.*** *(From H. Hamashima, M. Shibuta, and S. Itoh. 2011. Behavior of bubble in small water tank used for food processing using underwater shock wave.* Materials Science Forum *673 225–230).*

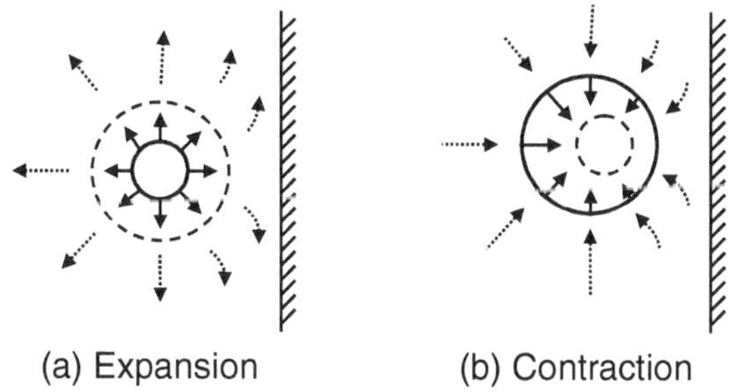

Fig. 1.10 ***Behavior of the bubble gas near the rigid body wall.*** *(From H. Hamashima, M. Shibuta, and S. Itoh. 2011. Behavior of bubble in small water tank used for food processing using underwater shock wave.* Materials Science Forum *673 225–230).*

designing the strength of marine structures such as ships. Furthermore, the behavior of the product gas is greatly affected by the boundary, affecting cases with a structure or not. As shown in Fig. 1.10, the product gas near the rigid wall loses its balance and tends to flow toward the center when contracted. After that, the flow on the opposite side of the wall increases, and the product gas flows toward the wall. In particular, it has been reported through experiments and numerical simulations that the product gas generated in a finite water tank settles toward the rigid wall due to the generation of a strong water flow [10].

1.3 High-rate processing of materials

For the high-rate processing of materials, the order of the velocity accelerating the materials ranges from 50 m/s to 1 km/s, depending on the processes. The velocity range is above the split Hopkinson pressure bar tests and is lower than the hypervelocity impact like space debris. Due to the short pressurizing time, an adiabatic heating effect exists and such an effect is more significant for the hypervelocity impact of space debris. Since space debris has quite large kinetic energy, the effect of the strength is not affected significantly, which results in inducing melting and/or vaporizing. For materials processing, it is necessary to recover the sample—without phase change and without defects—and thus the velocity to be employed should be limited.

Under a high strain rate of more than 100/s, the strain rate effect should be considered using Johnson–Cook and other constitutive equations considering the strain rate and the temperature effects for discussing high-rate deformation. But it is interesting that it is sometimes possible to treat the materials as point mass when discussing the terminal velocity of a flyer plate where the kinetic energy is higher than the deformation energy for the case of explosive welding [12]. During and after World War II, the equations of Gurney [13] and Deribas et al. [14], as well as other equations, were proposed based on the e/m ratio, where e is the explosive quantity and m is the mass of the flyer plate per unit area, which can estimate the terminal velocity of the flyer plate. After that, Deribas [12] suggested an equation to estimate the flyer plate velocity with flying distance that can describe the plate velocity change accelerated by the detonation gas behind the flyer plate. Currently, such velocity can be estimated using commercial codes like AUTODYN, based on a numerical approach [15,16]. For such numerical simulation, reliable property of the explosive should be known in advance of the simulation and is sometimes difficult to obtain. In addition, it is necessary to verify the accuracy of the numerical results based on measurement experiments.

So far, some related technologies for high-rate processing of materials have been developed, and these are demonstrated briefly in the following sections. The chapters mainly outline the processes because the major processes—explosive forming and explosive welding—are demonstrated in the following chapters in more in detail.

1.3.1 Explosive forming

Explosive forming was developed and actively used after World War II. There were some industrial demands, for example, to deform a large-sized dome of 10-ft. diameter, a pressure vessel bulkhead of 33-ft. diameter, a boiler roof panel, and others [17]. Because of the limited capacity of press machines at that time, explosive forming was employed for such large-sized forming of metal plates. The plates were formed through underwater explosion of an explosive [17]. The advantage of the process is sometimes explained based on the bubble pulse effect [18], and precise forming with the original die shape is expected by multi-time pressurization due to the bubble pulse phenomena, as mentioned earlier. The details of the current technology are discussed in Chapter 2.

1.3.2 Explosive cutting

There are two major procedures for cutting a metal plate using explosives. One method is multiple blanking of a pipe or a plate leading an intense underwater shock wave to a metal plate with a die machined with holes. Dynamic pressure may blank the plate making holes by a single shot of detonation of an explosive, and the technology is demonstrated in a book written by Ezra [17], who introduced a detonation cord in a steel pipe and succeeded to blank many holes simultaneously [17]. Fujita et al. [19] demonstrated a method to introduce high underwater pressure by detonating an explosive outside of a small stainless steel pipe and succeeded to blank five holes on the wall of a pipe. The methods are effective especially for drilling holes in the thin wall of a pipe and for blanking holes of noncircular shape.

Another cutting method is using linear- and conical-shaped charges [20]. Birkhoff et al. [21] discussed the process based on the inclination angle collision of two plates to form hypervelocity material flow ahead of the collision point. Such hypervelocity flow enables the cutting of thick metal plates and structures [20].

1.3.3 Explosive welding

Explosive welding is the most widely used technology in the field of high-rate processing of materials to join dissimilar metals with high bonding strength. The history and development of this technology have been documented by Crossland [22] and others. Therefore, only essential issues of the process are described here.

For explosive welding, it is necessary to accelerate a flyer plate to give a certain inclination angle β at the collision of two plates so as to induce a metal jet ahead of the collision point, as shown in Fig. 1.11, which is similar to the shaped charge. The metal jet cleans up the contamination and oxidation of the metal surfaces, and the following high pressure and an intense dynamic fluid-like flow cause a wavy interface that creates the tight bonding between the two components. The effect of heating is not significant for the joints obtained under optimized conditions. The formation of molten material or intermetallics under excessively energetic conditions

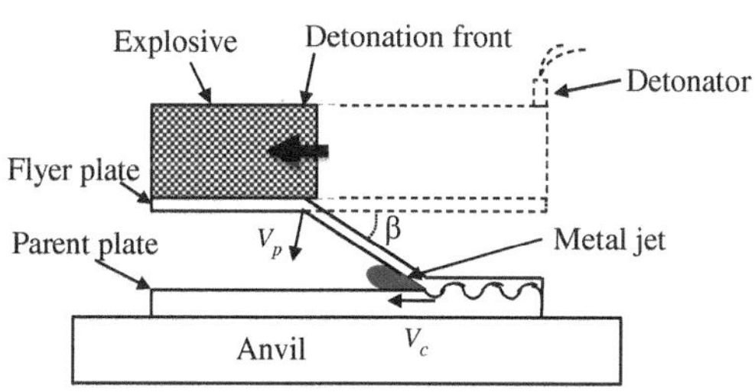

Fig. 1.11 ***Schematic illustration of an explosive welding process.***

causes separation of the joints. In addition, the welding is not achieved when the plate velocity Vp does not cause an intense deformation under lower energetic conditions. It is also necessary to decrease the horizontal collision point velocity Vc to lower than the sound velocity of the materials to be welded [23,24]. Such welding conditions are denoted by Wittman [25] and Deribas et al. [26] as the "welding window" for explosive welding. The details of the current industrial status and some recent achievements in designing the assembly, as well as the unique microstructural features, are discussed in Chapter 3.

1.3.4 Explosive hardening

Hardening a metal surface is possible by explosion treatment using explosives covering the metal plate (see the article by Deribas et al. [27]). Explosive hardening of high-manganese (Hadfield) steel was patented in 1955 [28]. Research and development work continued in the United States in the 1960s, and then the technology was spread to the former USSR, Japan, China, and other countries [27]. Production continues in Russia with the Novosibirsk Railroad Switch Plant producing 1 to 1.2 million pieces per year, and explosive hardening increases the lifetime of railroad switch frogs by 20% to 30% [27].

1.3.5 Shock synthesis

The use of explosive and other intense dynamic loading in a short period of time enables the synthesis of new materials through chemical reaction or phase transformation under conditions of extremely high pressure and high temperature. In the 1980s and 1990s, explosive compaction of mixed elements to synthesize various intermetallic compounds and composites was investigated [29–32], but it was not industrialized due to the difficulty of eliminating defects caused by the process.

The most important technology to be industrialized is the synthesis of diamond powders using explosives. The synthesized diamond powders are small and spherically shaped without edges, and are considered effective as abrasives for polishing. One method is using explosive compaction employing cylindrical geometry. A circular pipe filled with graphite and metal powders is dynamically compressed by the detonation of an explosive surrounding the circular pipe. The method was developed by DeCarli and Jamieson [33] and industrialized by DuPont [34].

Another method was developed in the former USSR [35] by detonating an explosive TNT and other high explosive in a big chamber. The detonation of an explosive exposed at high pressure and high temperature induces the decomposition of the explosive, and the carbon element is changed into nanometer-sized diamond powders recovered during the process. In Japan, Daicel Corporation started preliminary production in 2014, and their new market is for drag carriers, bioimaging, and diamond-based sensors [36]. It must be pointed out that the method was awarded an Ig Nobel Prize in 2012.

1.3.6 Others

Similar procedures using such intense dynamic loading are being considered for the softening of foods and/or sterilization depending upon the demands of food processing. This is because of the possibility to control the pressurizing condition relatively easily using underwater shock waves at relatively low pressure. The applied shock pressure in water can be changed easily just by changing the water distance, as mentioned in Section 1.2. Pioneering works were made by Solomon et al. [37] using a small amount of explosive followed by a method using electrical discharge [38,39]. The use of electrical discharge with sparks between electrodes [38] or metal wire explosion [39] in water is considered better due to the contamination of its residue compared with using explosive.

1.4 Conclusions

Due to the instantaneous phenomena in a short period, it is difficult to validate the process of high-rate processing of materials as to control the pressurizing condition with time. Recent developments in the equipment for measurements and the use of advanced numerical methods might facilitate its industrial application. The following chapters present some recent achievements made by the contributing authors for various cases of high-rate processing of materials.

References

[1] U.S. Army Materiel Command, General phenomenology. In: Engineering Design Handbook: Explosions in Air Part One, AMCP/06-181,U.S. Army Materiel Command, Alexandria, VA, 1974, pp. 1-3 –1-9.
[2] G.F. Kinney, K.J. Graham, Explosive Shocks in Air, Springer-Verlag, Berlin, Germany, 1985.
[3] C. N. Kingery and B. F. Pannill, Peak Overpressure vs Scaled Distance for TNT Surface Bursts (Hemispherical Charges). BRL Memorandum Report No. 1518. Ballistic Research Lab, Aberdeen Proving Ground, MD, 1964.
[4] W.E. Baker, Explosions in Air, University of Texas Press, Austin, 1973.

[5] H. Arai, R. Hiyoshi, Blast pressure measurement of TNT and commercial explosives above the ground [In Japanese.], Science and Technology of Energetic Materials 72 (1) (2011) 1–8.
[6] R.H. Cole, Underwater Explosion, Princeton University Press, Princeton, NJ, 1948.
[7] H.M. Schauer, Pressures and impulses from underwater explosions, In Proceedings of the 1st International Conference of the Center for High Energy Forming, Estes Park, CO, 1967 5 June 19--23-1-1 – 5-1-17.
[8] U.S. Army Materiel Command, Shock wave scaling. In: Engineering Design Handbook: Principles of Explosive Behavior, AMCP 706-180, U.S. Army Materiel Command, Alexandria, VA, 1972, pp. 13-33–13-37.
[9] K. Hokamoto, Challenges for unexplored materials processing technology [In Japanese.], Bulletin of the Japan Society for Technology of Plasticity 1 (4) (2018) 235–236.
[10] H. Hamashima, M. Shibuta, S. Itoh, Behavior of bubble in small water tank used for food processing using underwater shock wave, Materials Science Forum 673: (2011) 225–230.
[11] S. Itoh, Z. Liu, Y. Nadamitu, An investigation on the properties of underwater shock waves generated in underwater explosions of high explosives, Journal of Pressure Vessel Technology 119 (1997) 498–502.
[12] A. Deribas, Science of explosive welding: State of art, in: A. Chiba, S. Tanimura, K. Hokamoto (Eds.), Impact Engineering and Applications, Elsevier, Oxford, UK, 2001, pp. 527–534.
[13] R. Gurney. *The Initial Velocities of Fragments from Bombs, Shells and Grenades*. Report No. 405. Ballistic Research Laboratory, Aberdeen, MD, AII-36218, 1943.
[14] A.A. Deribas, V.M. Kudinov, F.I. Matveenkov, V.A. Simonov, Determination of the impact parameters of flat plates in explosive welding, Combustion, Explosion, and Shock Waves 3 (2) (1967) 182–186.
[15] S. Kumai, Impact joining of dissimilar metals [In Japanese.], Journal of the Japan Welding Society 87 (1) (2018) 48–56.
[16] X. Zeng, X. Li, X. Chen, X. Wang, H. Yan, Numerical and experimental studies on the explosive welding of plates with different initial strength, Welding in the World 63 (4) (2019) 967–974.
[17] A.A. Ezra, Principles and Practice of Explosive Metalworking, Industrial Newspapers Ltd., London, UK, 1973.
[18] R.H. Cole, Underwater Explosions, Princeton University Press, Princeton, NJ, 1948.
[19] M. Fujita, M. Mochihara, S. Nagano, E. Uematu, S. Itoh, A new method for making holes on the pipe wall by underwater shock waves, in: Z.R. Wang, H. Yuxin (Eds.), International Academic Publishers, Beijing, China, 1993, pp. 100–105.
[20] Wikipedia. 2020. Shaped charge. Retrieved April 6, 2020 from https://en.wikipedia.org/wiki/Shaped_charge.
[21] G. Birkhoff, D.P. MacDougall, E.M. Pugh, G.I. Taylor, Explosives with lined cavities, Journal of Applied Physics 19 (1948) 563.
[22] B. Crossland, Explosive Welding of Metals and Its Application, Clarendon Press, Oxford, UK, 1982.
[23] J.M. Walsh, R.G. Shreffler, F.J. Willig, Limiting conditions for jet formation in high velocity collisions, Journal of Applied Physics 24 (1953) 349–359.
[24] G.R. Cowan, A.H. Holzmann, Flow configuration in colliding plates: explosive bonding, Journal of Applied Physics 34 (1963) 928–939.
[25] R.H. Wittman, The influence of collision parameters on the strength and microstructure of an explosion welded aluminum alloy, In Proceedings of the 2nd International Symposium on Use of an Explosive Energy in Manufacturing Metallic Materials, Marianske Lasne, Czech Republic, 1973, pp. 153–168.

[26] A.A. Deribas, V.A. Simonov, I.D. Zakcharenko, Investigation of explosive welding parameters for arbitrary combinations of metals and alloys, In Proceedings of the 5th International Conference on High Energy Rate Fabrication, Denver, CO, June 24--26 1975, pp. 4.2.1–4.2.16.
[27] A.A. Deribas, A.A. Shtertser, E.E. Zubkov, Explosive hardening and its application in production of railroad switch frogs, Advanced Materials and Technologies 3 (2016) 3–16.
[28] N.A. McLeod, Method of hardening manganese steel. US Patent No. 2703297, (1955) March 1.
[29] Y. Horie, R.A. Graham, I.K. Simonsen, Observation on shock-synthesis of intermetallic compounds, in: L.E. Murr, K.P Staudhammer, M.A. Meyers (Eds.), Metallurgical Applications of Shock-Wave and High-Strain-Rate Phenomena, Marcel Dekker, New York, NY, 1985, pp. 1023–1035.
[30] L.H. Yu, M.A. Meyers, Shock synthesis and synthesis-assisted shock consolidation of suicides, Journal of Materials Science 26 (3) (1991) 601–611.
[31] N.N. Thadhani, S. Works, Shock-induced reaction synthesis (SRS) of nickel aluminides, Journal of Materials Research 7 (5) (1992) 1063–1075.
[32] B.R. Krueger, A.H. Mutz, T. Vreeland, Shock-induced and self-propagating high-temperature synthesis reactions in two powder mixtures: 5:3 atomic ratio Ti/Si and 1:1 atomic ratio Ni/Si, Metallurgical Transactions A 23 (1) (1992) 55–58.
[33] P.S. DeCarli, J.C. Jamieson, Formation of diamond by explosive shock, Science 133 (1961) 1821–1822.
[34] E. I du Pont Nemours and Company, *Netherland Patent Release No. 6506395*, (1965), November 22.
[35] A.I. Lymkin, E.A. Petrov, A.P. Ershov, G.V. Sakovutch, A.M. Staver, V.M. Titov, Production of diamond from explosives [In Russian.], Doklady Akademii Nauk SSSR 302 (1988) 611–613.
[36] T. Mahiko, Introduction of detonation nanodiamond [In Japanese.], Explosion 29 (2) (2019) 98–100.
[37] M.B. Solomon, J.B. Long, J.S. Eastridge, The Hydrodyne: a new process to improve beef tenderness, Journal of Animal Science 75 (6) (1997) 1534–1537.
[38] U.M. Alvarez, A.M. Loske, E. Castaño-Tostado, F.E. Prieto, Inactivation of *Escherichia coli* O157:H7, *Salmonella typhimurium* and *Listeria monocytogenes* by underwater shock waves, Innovative Food Science and Emerging Technologies 5 (4) (2004) 459–463.
[39] A. Yasuda, E. Kuraya, A. Touyama, O. Higa, K. Hokamoto, S. Itoh, Underwater shock-wave pretreatment process for improving carotenoid content and yield of extracted carrot (*Daucus carota* L.) juice, Journal of Food Engineering 211 (2017) 15–21.

CHAPTER 2

Explosive forming

Hirofumi Iyama[a], Masatoshi Nishi[a], Shigeru Tanaka[b]
[a]National Institute of Technology, Kumamoto College, Kumamoto, Japan
[b]Kumamoto University, Japan

2.1 Introduction

Explosive forming has been used in the manufacture of large metal parts since the 1950s. This method presents several advantages: (1) unlike general press processing, the workpiece can be processed by using a female die only; (2) since the material undergoes high-speed deformation, better elongations are achieved; (3) the spring back effect is small owing to sufficient plastic deformation; and (4) since the material to be formed collides with the die surface at a high speed, the shape of the die surface is accurately transferred.

In the conventional explosive forming method, the explosive is detonated underwater, and the explosion energy is converted into hydraulic energy, which acts on the forming material. This method has been widely applied in various forms, such as metal plate forming, drawing, material surface coining, and drawing or extrusion of solid or hollow rods. Apart from the metal plate forming method based on the underwater shock wave generated by an underwater explosion, high-voltage discharge and electromagnetic forming methods that employ electromagnetic induction or Lorentz forces have also been developed.

In this chapter, the explosive forming methods developed to date are introduced, including explosive forming using a die and a closed pressure vessel, free forming and its numerical simulation, and spherical product forming without a die; application examples of explosive forming to realize fine processing are also presented.

2.2 Explosive forming of a metal plate

2.2.1 Conventional explosive forming method using a metal die

Explosive forming has been performed since the 1950s, and the forming of large dome parts was performed flourishingly from the 1950s to the 1960s [1]; the forming of the heads of SEALAB [2] is a representative example

Explosion, Shock-Wave and High-Strain-Rate Phenomena of Advanced Materials
DOI: 10.1016/C2019-0-03954-4

of this phenomenon. Rocket head explosive forming [3] was an important production method employed in the field of aerospace engineering. Some fine parts and metal base plates for teeth were also produced by explosive forming.

Fig. 2.1 shows the setup of the conventional explosive forming method. First, the blank is placed on the die; then, the blank and die are submerged in water with an airtight space between them. Finally, the explosive is set underwater at a suitable distance from the blank. When the explosive is detonated underwater, a shock wave is generated and the shock pressure acts on the surface of the blank. When a metal plate is deformed by an underwater shock wave, its deformation velocity is significantly high; if air remains between the blank and the die, it does not escape, experiencing an adiabatic compression and high pressure, thereby disturbing the deformation. Therefore, a vacuum should be generated in the space between the blank and the die beforehand. The underwater shock wave that acts on the blank has a significantly high peak pressure and a short pulse duration. Consequently, it generates a temporary displacement velocity in the blank. Since the shock pressure drastically decreases afterward, the water particle velocity is lower than the displacement velocity of the blank, which generates cavitation on the surface of the blank. When the rarefied expansion wave originating from cavitation reaches the high-pressure gas generated by the explosion, the expansion motion of the gas and water rapidly drives

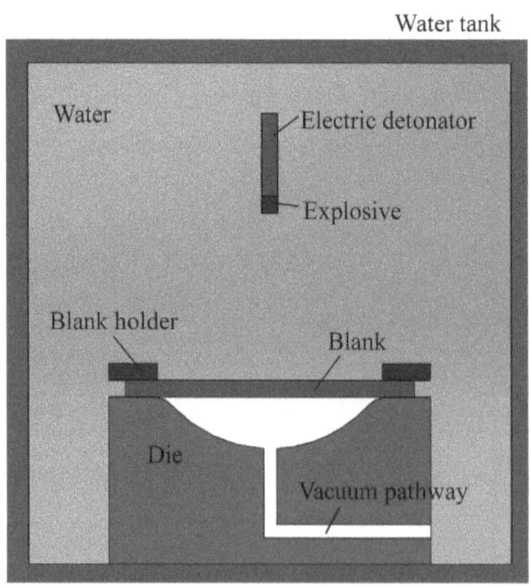

Fig. 2.1 ***Schematic of the setup of the conventional explosive forming method.*** *(From J. S. Rinhert and J. Person.* Explosive Working of Materials. *Pergamon Press, New York, NY, (1963), 194).*

them into the cavitation region. Therefore, the blank is subjected to water hammer action, which further promotes the deformation of the blank. The deformation process seems to proceed in two stages: deformation is first caused by the underwater shock wave and then by the water hammer accompanied by gas expansion. The stage with the highest contribution to deformation is determined by factors such as the distance between the explosive and the blank, and the ratio of the water level and diameter of the vessel to the size of the blank. The water hammer effect is reduced in cases of significant distance between the explosive and the blank, low water level, and small vessel diameter.

As an example of explosive forming using a metal die, a copper plate forming case study is now presented. Fig. 2.2 shows the experimental setup used, which comprised a pressure vessel and a die. The material of the metal die was carbon steel, and the die was held on a metal container. A circular copper plate with a diameter of 100 mm and a thickness of 1 mm was placed on the die. The maximum depth of the conical die was 23.8 mm, and the inclined angle was 35 degrees. The die presented a 1-mm-diameter hole at its bottom, which was connected to a vacuum pump located outside the

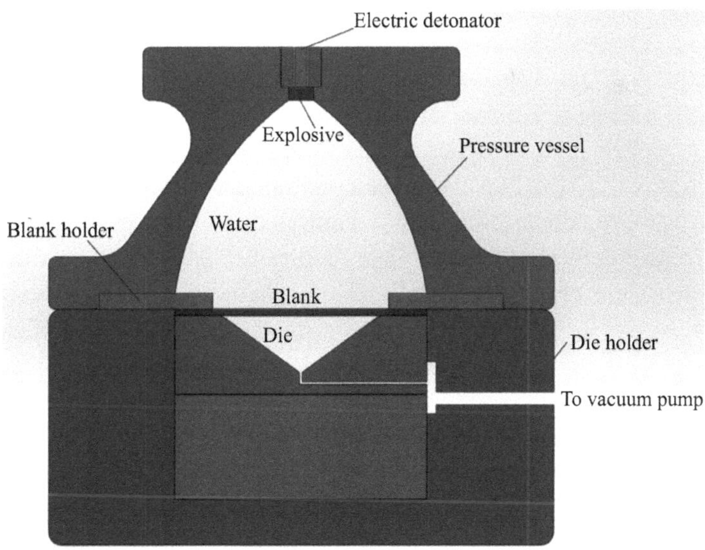

Fig. 2.2 ***Experimental setup for explosive forming using a pressure vessel.*** *(From M. Fujita, Y. Ishigori, M. Mochihara, H. Osaka, R. Zhang, and H. Iyama. Effects of pressure distribution and boundary condition on deformation mechanism of a circular metal plate subjects to shock pressure. [In Japanese.]* Journal of the Japan Society for Technology of Plasticity *38 (435) (1997) 331–336).*

Fig. 2.3 ***A half-cut copper plate deformed using a conical die.*** *(Photo provided by Y. Ishigori).*

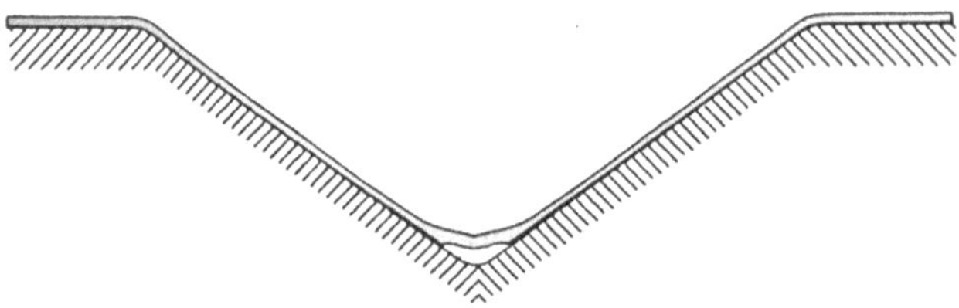

Fig. 2.4 ***Cross section of the copper plate and conical die, obtained using a 3D measurement system.*** *(Photo provided by Y. Ishigori).*

water tank. A dome-shaped pressure vessel was used. A high explosive—SEP [4]—with a mass of 1 g was set at the top of the pressure vessel. The peak pressure applied to the copper plate was approximately 100 MPa, and its duration was 40 μs [5]. Fig. 2.3 shows a photograph of the experiment result, with a half-cut copper plate and a conical die. The shape measurement results of the deformed copper plate (before being cut) and of the metal die, which were obtained using a 3D measurement system, are shown in Fig. 2.4. As the shock pressure acting on the copper plate was sufficient in the experiment, the shapes of the copper plate and the metal die agreed well.

2.2.2 Free forming of an aluminum plate

In the free forming method [6], the die used presents a cavity at its central part. In this section, an aluminum plate free forming case study is introduced. As aluminum alloys have a limited forming performance compared with conventional steel-based materials, the formability (i.e., the forming limit) of aluminum plates in the free forming process has been investigated [7].

The experimental equipment used for aluminum plate free forming is shown in Fig. 2.5. A 1-mm-thick aluminum plate made of A5052-O alloy was

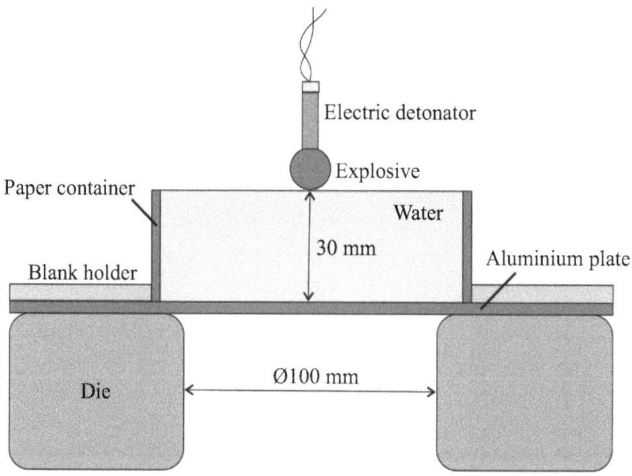

Fig. 2.5 ***Schematic of aluminum plate explosive forming using a paper container.*** *(From H. Iyama and S. Itoh. Study on explosive forming of aluminum alloy.* International Journal of Multiphysics *4 (4) (2010) 341–349).*

placed between the die and the blank holder. The SEP explosive (10 g), which was connected to the tip of an electric detonator, was placed on the water surface, at a 30-mm distance from the aluminum plate. It should be noted that an open container made of paper is used in this method. The detonation velocity of the SEP explosive was 6970 m/s, and the detonation pressure was 15.9 GPa [4]. Fig. 2.6 shows photographs of aluminum alloy plates obtained by explosive forming and static press forming. In the case of static press forming, the radius of the die shoulder was 4 mm, the clearance between the punch and the die was 2.5 mm, and the punching cycle was 35 cycles/min [8]. The aluminum plate was found to bulge well; under the aforementioned experimental conditions, the maximum bulge depth was approximately 39 mm.

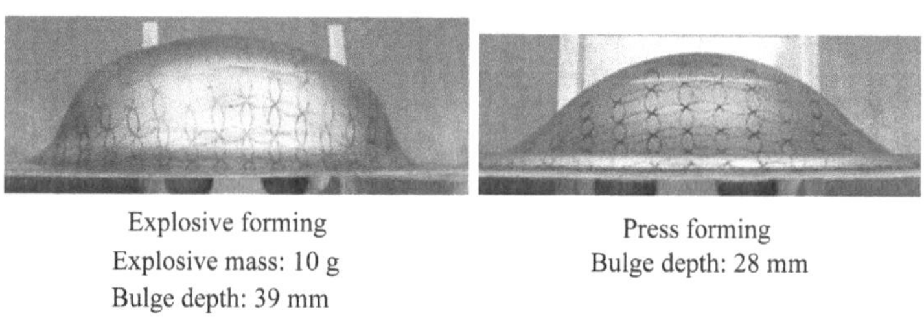

Fig. 2.6 ***Comparison between specimens obtained by aluminum plate explosive forming and static press forming.*** *(From H. Iyama and S. Itoh. Study on explosive forming of aluminum alloy.* International Journal of Multiphysics *4 (4) (2010) 341–349).*

2.2.3 Numerical simulation of explosive forming

Numerical simulations have been performed to determine the pressure and propagation of the shock wave generated during underwater explosive detonation and to investigate the deformation process of the metal plate during explosive forming. The simulation model described herein was created based on Fig. 2.5; however, in the numerical simulation, the paper vessel was ignored and the surrounding wall of the water column was regarded as a free surface. This numerical simulation was performed using a finite difference method based on a Lagrangian scheme [8]. The water pressure calculation was based on the Mie– Grüneisen equation of state [9], as shown in Eq. (2.1):

$$P_w = \frac{\rho_0 c_0^{\,2}\eta}{(1-s\eta)^2}\left[1-\frac{\Gamma_0\eta}{2}\right]+\Gamma_0\rho_0 e, \tag{2.1}$$

where P_w is the water pressure (Pa); ρ_0 is the initial density of water (=1000 kg/m^3); C_0 is the sound velocity in water (=1500 m/s); $\eta = 1 - \rho_0/\rho$; ρ is the density of water (kg/m^3); Γ_0 and s are the Grüneisen parameters of water (Γ_0 = 1.65 and s = 1.79); and e is the internal energy (J).

The pressure generated by the explosion is calculated by the Jones–Wilkins–Lee (JWL) equation of state [10], as shown in Eq. (2.2):

$$P_g = A\left[1-\frac{\omega}{VR_1}\right]\exp(-R_1V)+B\left[1-\frac{\omega}{VR_2}\right]\exp(-R_2V)+\frac{\omega e\rho_0}{V} \tag{2.2}$$

where P_g is the pressure of the explosion gas (GPa); V is the specific volume (m^3/kg); and A, B, R_1, R_2, and ω are the JWL parameters of the SEP explosive.

The JWL parameters of the SEP explosive are shown in Table 2.1.

The constitutive equation of the aluminum plate was applied to the following simplified Johnson–Cook model (Eq. (2.3)) [12]:

$$\sigma_y = \left[A+B(\varepsilon_p)^n\right]\left[1+Cln\left(\frac{\dot{\varepsilon}_p}{\dot{\varepsilon}_0}\right)\right], \tag{2.3}$$

where σ_y is the equivalent yield stress (MPa); ε_p is the plastic strain; $\dot{\varepsilon}_p$ is the equivalent plastic strain rate (1/s); $\dot{\varepsilon}_0$ is the reference strain rate [1/s]; and A, B, C, and n are the material parameters.

Table 2.1 JWL parameters of the SEP explosive [11].

A (GPa)	B (GPa)	R_1 (-)	R_2 (-)	ω (-)
365	2.31	4.30	1.10	0.28

From S. Itoh, Z. Liu, and Y. Nadamitsu. An investigation on the properties of underwater shock waves generated in underwater explosions of high explosives. Journal of Pressure Vessel Technology 119 (4) (1997) 498–502.

Table 2.2 Material parameters of A5052 alloy [13].

A (MPa)	*B* (MPa)	*C* (-)	*n* (-)
199	8.52	310	0.367

From S. A. Hosseini Kordkheili, M. M. Ashrafian, and H. Toozandehjani. A rate-dependent constitutive equation for 5052 aluminum diaphragms. Materials and Design 60 (2014) 13–20.

The values of the aforementioned material parameter for A5052 are shown in Table 2.2.

The underwater shock wave propagation process obtained in the simulation is shown in Fig. 2.7. When the time is 0 μs, the blackened area shows the initial detonation position; it can be seen that the underwater shock wave spreads spherically after the detonation. At approximately 15 μs, the shock wave reaches the center of the aluminum plate and spreads across the plate surface. Since the top and sides of the water column were modeled as free surfaces, no reflected shock waves are observed, and the pressure is rapidly attenuated. Fig. 2.8 shows the pressure history at the center of the aluminum plate surface and at 10-mm-distance intervals from it; these pressure values include the pressure reflected from the surface of the aluminum plate. High pressures exist up to approximately 30 μs, the highest pressure value is located at the center of the plate, and the peak pressure values decrease as the distance from the plate center increases.

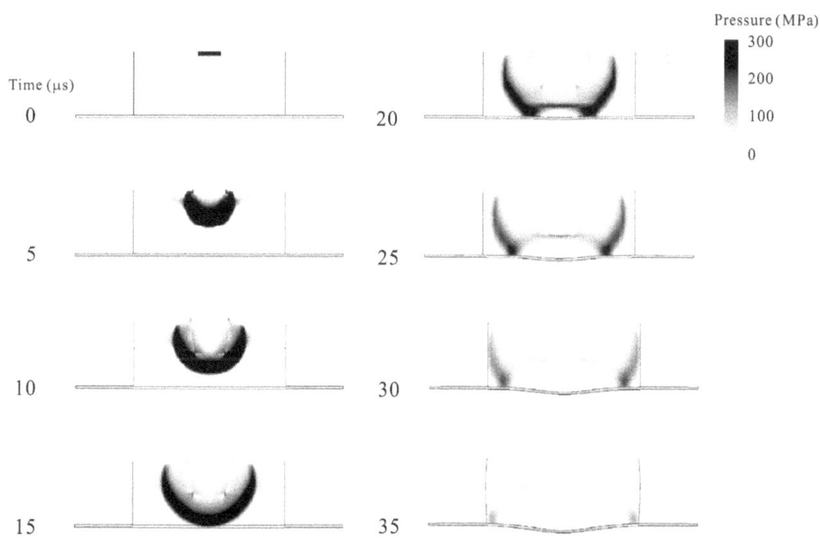

Fig. 2.7 ***Propagation process of the underwater shock wave generated by SEP explosive detonation via a numerical simulation.*** *(From H. Iyama and S. Itoh. Study on explosive forming of aluminum alloy.* International Journal of Multiphysics *4 (4) (2010) 341–349).*

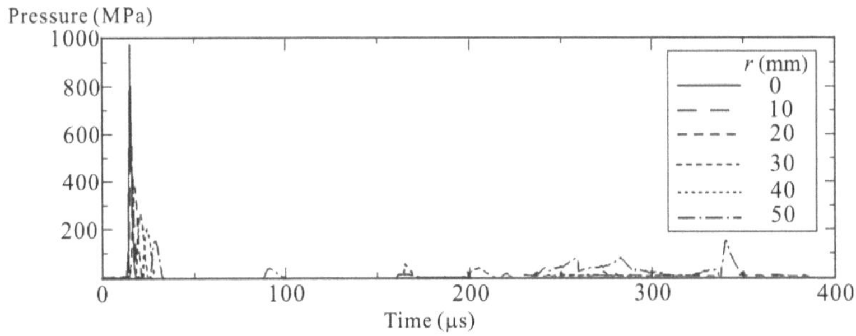

Fig. 2.8 ***Pressure profile in the water on the aluminum plate at the center of the plate and at 10-mm-distance intervals from it (r: radial position from the center).*** *(From H. Iyama and S. Itoh. Study on explosive forming of aluminum alloy.* International Journal of Multiphysics *4 (4) (2010) 341–349).*

Fig. 2.9 shows the deformation process of the aluminum plate from 0 to 400 μs. Between 20 and 40 μs, deformation appears at the central part of the plate. The plastic bending wave apparently approaches the die shoulder at 60 μs, and the plate bends down slightly around the die

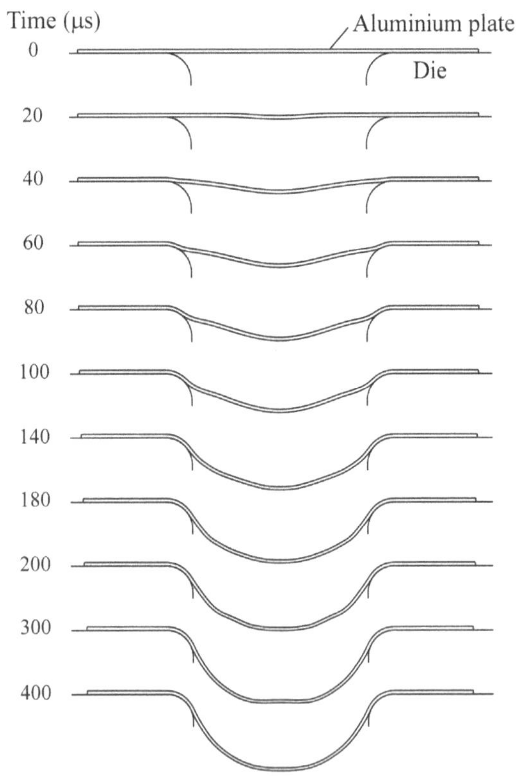

Fig. 2.9 ***Deformation process of the aluminum plate from 0 to 400 μs.*** *(From H. Iyama and S. Itoh. Study on explosive forming of aluminum alloy.* International Journal of Multiphysics *4 (4) (2010) 341–349).*

opening as the deformation process progresses. The bulge at the die shoulder continues to deepen toward the center, where the bulge depth is highest; the aluminum plate section over the die opening adopts a hemispherical shape at the end of the process. The bulge depth at the center of the plate at 400 μs, when the deformation was almost completed, was approximately 41 mm. The bulge depth determined experimentally was 39 mm.

2.3 Non-die explosive forming of spheres

The non-die explosive forming method can be used to produce spheres [14]. This method is significantly different from the die-based forming and free forming methods described in previous sections, because no dies or large presses are used in the forming operation. Instead, a multicone structure needs to be prewelded to form a sealed container, as shown in Fig. 2.10A. In non-die explosive forming, this prewelded structure is formed into a sphere owing to its "self-controlled" deformation behavior. Compared with the conventional explosive forming process described in Section 2.2.1, the non-die explosive forming method presents the following advantages: (1) die fabrication costs are eliminated, (2) the energy released from the explosive can be effectively used, and (3) the intensity of the generated noise or vibration is highly reduced because the explosive is detonated inside a sealed water-filled quasispherical container. It should be noted, nonetheless, that problems related to the strength of the prewelded regions unavoidably arise in this process. Moreover, in this method, stress analysis becomes more important than in conventional explosive forming. Briefly, the fundamental principle of the non-die explosive forming

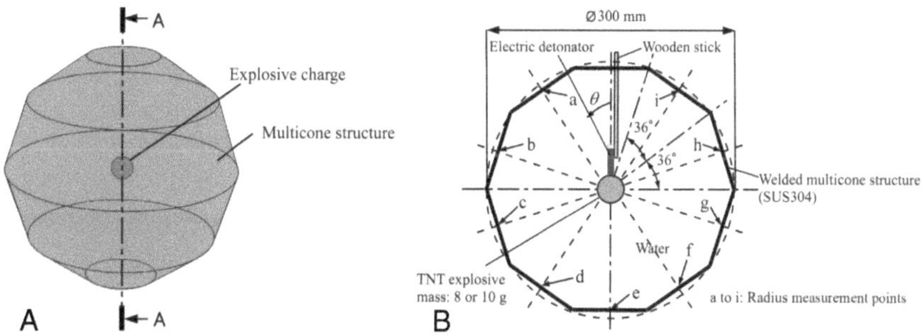

Fig. 2.10 ***Schematic of the non-die explosive forming process.*** (A) 3D drawing. (B) A–A section of the 3D drawing.

Fig. 2.11 ***The spherical shell before (A) and after (B) forming.***

method is as follows: the maximum stress generated in the welded region should exceed the strength of this area but not the fracture limit of the plate.

Fig. 2.10B shows the experimental setup used in the sphere forming case study described herein. Stainless steel plates of 2-mm thickness were welded to form a quasispherical shell. The rotation angle of the edges of the shell was 36 degrees. The shell was assembled so that it adopted a 300-mm-diameter spherical shape after forming. An 8- or 10-g spherical TNT explosive and an electric detonator were placed in the center of the shell using a wooden stick. After and before forming, the radius of the shell was measured at nine points, denoted by lowercase letters from "a" to "i"; these measurement points were located in the middle of the shell, as shown in Fig. 2.10B. The θ angle was set counterclockwise from the center of the top of the quasispherical shell. Fig. 2.11 shows photographs of the shell before and after forming for the case in which a 10-g explosive mass was used; it can be seen that the quasispherical shell deformed to a spherical shell. Fig. 2.12 shows the radius measurement results at points "a" to "i" defined in Fig. 2.10B, before and after forming. In this figure, the horizontal axis is θ, and the radius error at each measurement point is plotted. These results show that the forming deviation is significantly small (less than 1%), except for the top area of the spherical shell. In the case with 10 g of explosive, the original shell became more spherical than in the case with 8 g of explosive.

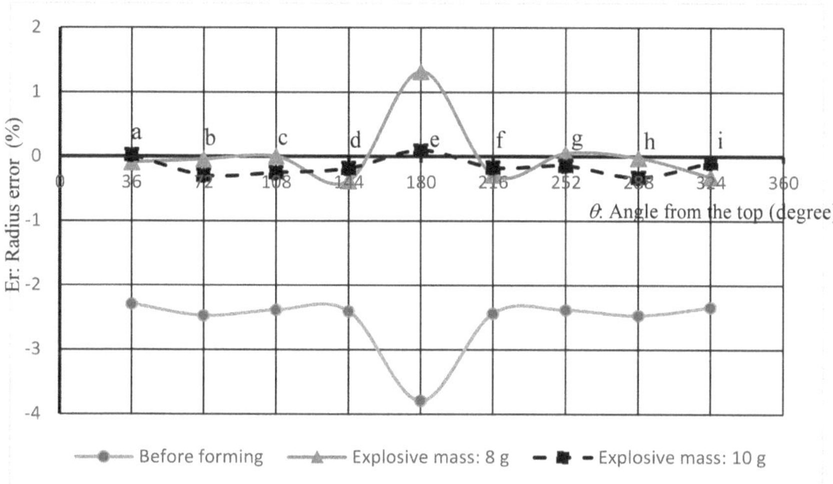

Fig. 2.12 ***Experimental results of non-die forming using different amounts of explosive charge.***

2.4 Microforming using an underwater shock wave

2.4.1 Microforming via hydro spark forming

Hydro spark forming utilizes an impulsive hydraulic pressure wave generated by a high-voltage electrical discharge [15–17]. The electrical energy stored in a capacitor bank is discharged through a pair of electrodes bridged with thin aluminum or copper wire. When an electric current flows through this thin wire, the wire is instantaneously melted and vaporized, which generates an underwater shock wave. After this phenomenon occurs, the produced gas bubbles expand rapidly, thereby causing a dynamic hydraulic pressure inside the vessel. The magnitude of the generated hydraulic pressure shocks is determined by the amount of electrical energy employed, which is proportional to the square of the voltage and the capacitance of the capacitor. Hasebe et al. [17] attempted to develop a simple yet robust technique for the microforming of thin foils. The developed hydro spark forming method was used for submillimeter-scale bulging and punching and for millimeter-scale surface embossment. They reported a major contribution of the former to blank deformation in an open-type apparatus. Since the hydro spark forming technique produces significantly high strain rates and high-pressure conditions, it presents many unique features compared with conventional explosive forming methods, such as the capability to process difficult-to-form materials and to obtain products with complex shapes. As even minute flaws remaining on the die surface are accurately transferred to the blank, this method is expected to be suitable for a variety of microforming applications. For example, it is possible to shape a 25-μm-thick aluminum plate if it presents a diameter and height of approximately 300 and 150 μm, respectively. In addition, coins with a jaggedness below 100 μm and integrated circuit chips chips and holograms with a thickness below 10 μm can be formed [18].

2.4.2 Application to artworks

Artwork products are highly interesting applications of explosive forming. The explosive forming method able to achieve sufficient plastic deformation has been known to be suitable for fine forming for more than 40 years. Fujita et al. [19] have exploited this characteristic and developed artworks by explosive forming in collaboration with Asahi Kasei Corporation and Nikko Gika Company Ltd.; these artworks were designed as wall panels for buildings. A highly interesting feature of this technique is that paper, gypsum, sand, resin, and rubber can be used as die materials [20,21]. In explosive forming, the metal plate subjected to the high pressure of the underwater shock wave travels at a high velocity and collides with the die; this causes the sufficient and plastic deformation of the metal plate and accurately transfers the die shape to the plate, with only minor imperfections. Hence, even if a nonmetallic die is used, the metal plate can be easily deformed. When paper or rubber is subjected to vertical pressure, the material expands laterally and deforms significantly with a relatively low pressure. However, if these materials are subjected to an impact load, the metal plate presses the die entirely; therefore, the metal plate cannot spread laterally and the die is compressed vertically, which increases its density. As this causes a large stress, the metal plate experiences a large deformation with a slight compression, and the motion of the metal plate ceases rapidly. When a metal die and a paper die are compared, it seems that there is a difference of 10% or less in the amount of deformation experienced. With such a difference, the appearance of the product is almost comparable.

Fig. 2.13 shows two examples of artworks obtained via explosion forming. Fig. 2.13A shows a work obtained by forming a 0.3-mm-thick annealed copper plate. After the forming process, the copper plate was corroded with an acid solution to produce shading. To obtain this artwork, a pattern was printed on medical imaging film and cut out to form a die. A simplified artwork explosive forming method based on household vacuum packing was developed. A lotus leaf was placed on a metal base plate, and a 50-μm-thick aluminum plate was placed over the leaf. Then, the aluminum

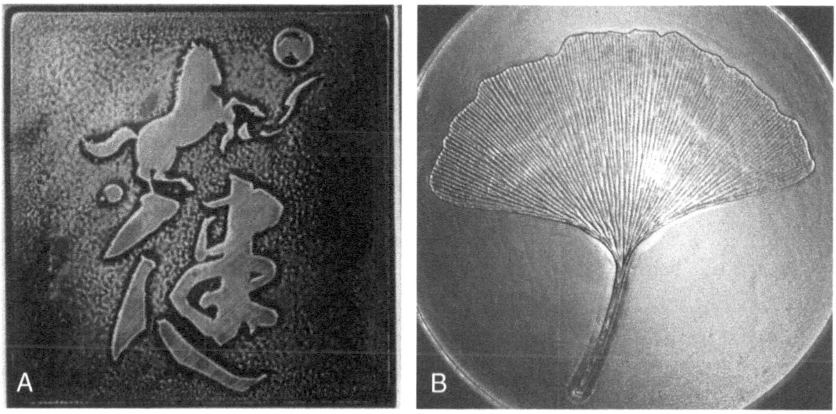

Fig. 2.13 ***Examples of artworks produced via explosive forming.*** (A) Artwork obtained using a paper and sand die. (B) Artwork obtained using a lotus leaf die and a high-voltage electrical discharge. *(Photo (A) provided by M. Fujita).*

plate was deformed by a shock wave generated by underwater thin metal wire high-voltage discharge. As observed in Fig. 2.13B, the leaf contour and veins were accurately transferred to the copper plate.

2.4.3 Micropunching

Currently, there is an increasing demand for the miniaturization of devices in the high-technology industry. Therefore, in the punching technique, small hole diameters and high processing precision are required. Thin metal foil micropunching has garnered significant attention in the industry [22]. In general, a micro die and punch made of sufficiently hard materials, such as cemented carbide or high-speed steel, are required in this method, and holes of various shapes and in the order of tens to hundreds of micrometers can be punched. Recently, a laser shock-based micropunching technique was developed [23]; however, it presents limitations related to the size of the punching area, which is in the order of the laser beam size. Therefore, a large area cannot be processed in a single step, which leads to long times to manufacture products with many holes. This size limitation can be easily overcome by using underwater shock waves generated by high explosive detonation or electrical discharge explosions. To date, punching technology to process large areas has been developed based on shock waves generated by the detonation of explosives. Since the intensity of the shock wave propagating in water attenuates with distance, the shock pressure can be controlled by changing the distance between the blank and the explosive and by changing the explosive type. Explosive forming was used to accurately transfer the complex shape and microrelief of the die to a copper plate. The explosive forming method can realize fine forming and punching. Mochihara et al. [24] were able to produce a single 10-mm-diameter hole in a 1.6-mm-thick copper plate by using the shock wave generated by underwater explosive detonation. In addition, research on metal plate or tube punching using the shock wave generated by underwater explosive detonation has also been conducted [25-28]. However, no other submillimeter punching research has been conducted to date. Tanaka et al. [29] proposed a new approach for simultaneously forming multiple micropores across a thin metal foil wide area by using an underwater shock wave. In their study, explosives were used to punch multiple holes simultaneously in an 80 × 45 mm metal sheet, which was significantly wider than the sheets used in laser punching.

Fig. 2.14A shows the setup for a small-scale experiment using a cylindrical explosive. The cylindrical explosive, which had a diameter of 18 mm, a height of 10 mm, and a mass of 3 g, was placed above the metal sheet. High-power SEP based on PETN was used as the explosive. The metal die used presented multiple rows of 0.3-mm-diameter holes made in a 0.3-mm-thick SUS304 stainless steel plate via chemical etching; the hole

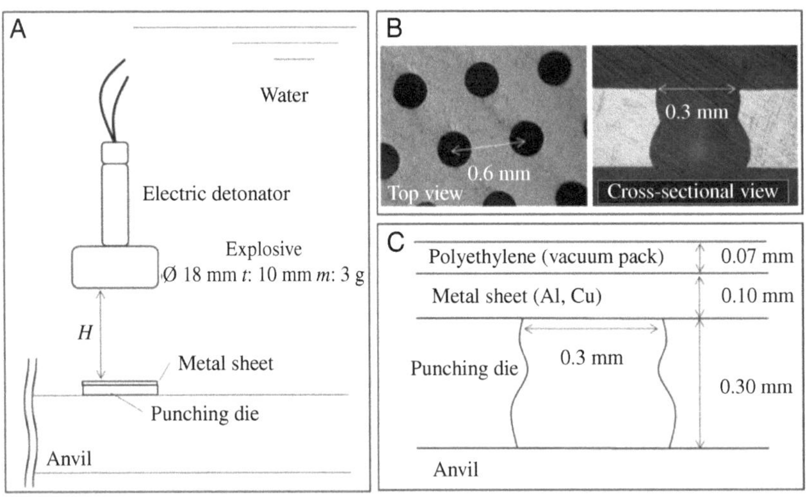

Fig. 2.14 ***Underwater shock wave–based micropunching***. (A) Schematic of the experimental setup. (B) Punching die made of SUS304 alloy with holes obtained using electrochemical etching. (C) Layout of the punching die and metal sheet. *(From S. Tanaka, I. Bataev, M. Nishi, I. Balagansky, and K. Hokamoto. Micropunching large-area metal sheets using underwater shock wave: experimental study and numerical simulation.* International Journal of Machine Tools and Manufacture *147 (2019) 103457).*

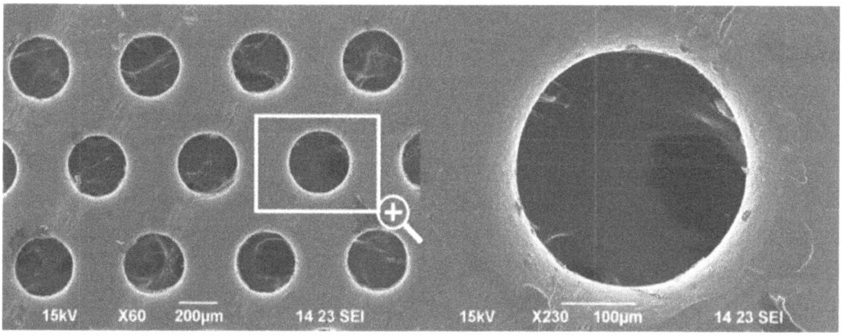

Fig. 2.15 ***Top view of the punched holes and a magnified image of a single hole.*** *(From S. Tanaka, I. Bataev, M. Nishi, I. Balagansky, and K. Hokamoto. Micropunching large-area metal sheets using underwater shock wave: experimental study and numerical simulation. International Journal of Machine Tools and Manufacture 147 (2019) 103457).*

spacing was 0.6 mm. Fig. 2.14B shows the hole pattern obtained and a cross-sectional view of the obtained holes. The punching material used was pure aluminum (A1100) with a foil thickness of 0.1 mm. The sheet was fixed to the surface of the punching die and the assembly was placed in a polyethylene container and vacuum packed, as shown in Fig. 2.14C. Then, the entire assembly was submerged in water.

Fig. 2.15 shows scanning electron microscope photographs of the holes formed throughout the aluminum foil when the distance between the explosive and the metal plate (*H*) was 40 mm, as shown in Fig. 2.14A. No unpunched holes, damage to the aluminum foil between the holes, or other defects were observed. Since A1100 alloy exhibits a relatively high ductility, burrs were formed owing to the significant elongation at the edge of the punching die. The hole shape was similar to that of the holes punched using a laser shock wave [30]. The comparison between underwater shock wave punching and laser-induced shock wave punching is highly interesting. In general, for punching relatively small holes, the material thickness should be reduced. The shock wave generated by explosive detonation has a significantly longer compression pulse duration than that of a normal laser-induced shock wave. However, with the underwater shock wave–based method, it was possible to punch a plate with a thickness of 100 μm, even when the ratio of cross-sectional area to hole perimeter was 0.075.

2.5 Conclusions

Different explosive forming methods and application examples were presented in this chapter. First, not only basic metal sheet forming but also free forming and its numerical simulation were introduced. Then, the non-die forming method for spherical shells was described. In addition, a microforming application example that was not based on the conventional explosive forming method was introduced. As highly interesting methods, the production method of artworks using a die made of paper or leaves and the fine punching process, which has been garnering significant attention in recent years, were presented. Based on the results presented herein, the explosive forming method is expected to be further developed to produce new technologies.

References

[1] D.M. Shuford, The chemical milling of explosive formed 10-feet diameter aluminum domes, In Proceedings of the 1st International Conference of the Center for High Energy Forming, Denver, CO, June (19–23) 1967, pp. 1-3-6–1-3-31 .
[2] S. Giannoccolo, N. Architect, Explosive forming of 12-feet diameter end closures for SEALAB 2, In Proceedings of the 1st International Conference of the Center for High Energy Forming, Denver, CO, June (19–23) 1967, pp. 1-3-6–1-3-35.
[3] J.S. Rinhert, J. Person, Explosive Working of Materials, Pergamon Press, New York, NY, 1963, pp. 194.
[4] S. Itoh, Y. Nadamitsu, A. Kira, S. Nagano, M. Fujita, T. Honda, Fundamental characteristics of underwater shock wave due to underwater explosion of high explosives [In Japanese.], Transactions of the Japan Society of Mechanical Engineers Series B 62 (601) (1996) 3278–3283.
[5] M. Fujita, Y. Ishigori, M. Mochihara, H. Osaka, R. Zhang, H. Iyama, Effects of pressure distribution and boundary condition on deformation mechanism of a circular metal plate subjects to shock pressure [In Japanese.], Journal of the Japan Society for Technology of Plasticity 38 (435) (1997) 331–336.
[6] M. Fujita, Y. Ishigori, M. Mochihara, H. Osaka, S. Itoh, Numerical analysis of a circular plate clamped at the periphery under the shock loading by employing the effects of strain and strain-rate hardening, and thickness [In Japanese.], Journal of the Japan Society for Technology of Plasticity 36 (412) (1995) 541–546.
[7] H. Iyama, S. Itoh, Study on explosive forming of aluminum alloy, International Journal of Multiphysics 4 (4) (2010) 341–349.
[8] T. Takeshi, H. Iyama, S. Itoh, Study of free forming of aluminum alloy using underwater shock wave (I): optimum conditions of aluminum alloy forming [In Japanese.], Journal of the Japan Explosive Society 66 (3) (2005) 301–306.
[9] R.G. McQueen, S.P. Marsh, J.W. Taylor, J.N. Fritz, The equation of state of solids from shock wave studies, in: R. Kinslow (Ed.), High-Velocity-Impact Phenomena, Academic Press, New York, NY, 1970, pp. 293–417.
[10] E. Lee, M. Finger, W. Collins, JWL Equation of State Coefficients for High Explosives, Report UCID-16189, Lawrence Livermore National Laboratory, Livermore, CA, 1973.
[11] S. Itoh, Z. Liu, Y. Nadamitsu, An investigation on the properties of underwater shock waves generated in underwater explosions of high explosives, Journal of Pressure Vessel Technology 119 (4) (1997) 498–502.
[12] G.R. Johnson, W.H. Cook, A constitutive model and data for metals subjected to large strains, high strain rates and high temperatures, In Proceedings of the 7th International Symposium on Ballistics, The Hague, Netherlands, April (19–21), 1983, pp. 541–547.

[13] S.A.Hosseini Kordkheili, M.M. Ashrafian, H. Toozandehjani, A rate-dependent constitutive equation for 5052 aluminum diaphragms, Materials and Design 60 (2014) 13–20.
[14] R. Zhang and T. S. Zhang. Non-die explosive forming of spherical pressure vessel. Journal of Material Processing Technology 41 (1994) 341–347.
[15] T. Hasebe, Y. Imaida, M. Yokoi, Flat-heated cylindrical deep drawing of aluminum foil by hydro spark forming method, dynamic plasticity and structural behaviors, In Proceedings of the International Conference on Plasticity, University of Osaka Prefecture, Sakai, Osaka, Japan, 1995, pp. 991.
[16] T. Hasebe, Y. Imaida, High strain rate deep drawing by hydro-spark forming method, In Proceedings of the 3rd International Conference (Mumisheet '96), Dearborn, MI, September 29–October 3 1996, pp. 372.
[17] T. Hasebe, Y. Takenaga, H. Kakimoto, Y. Imaida, High strain rate forming using underwater shock wave focusing technique, Journal of Materials Processing Technology, 85 (1998) 194.
[18] T. Hasebe, Y. Imaida, On micro forming by hydro spark forming method, Material Science Forum 566 (2008) 167–172.
[19] M. Fujita, Y. Ishigori, S. Nagano, N. Kimura, S. Itoh, Explosive precision of fine arts using regulated underwater shock wave, Advances in Engineering Plasticity and Its Application (1993) 1007–1013.
[20] K. Shiramoto, A. Mori, T. Watanabe, K. Hisanabe, M. Fujita, Explosive forming techniques for making a copper plate relief, Science and Technology of Energetic Materials 73 (4) (2012) 104–108.
[21] K. Shiramoto, T. Watanabe, A. Mizuno, H. Iyama, M. Fujita, A method of explosive forming process for making a palm and a back of human hand, Materials Science Forum 767 (2013) 188–191.
[22] C. Diver, J. Atkinson, H.J. Helml, L. Li, Micro-EDM drilling of tapered holes for industrial applications, Journal of Material Processing Technology 149 (1-3) (2004) 296–303.
[23] L. Song, X. Zhang, Y. Zhang, H. Li, Z. Ji, C. Zheng, Shortening post-processing and improving forming quality of holes in laser shock punching with the aid of silicone rubber, Optics and Laser Technology 106 (2018) 442–450.
[24] M. Mochihara, K. Hokamoto, M. Fujita, Shearing of a metal plate using underwater shock wave and the mechanism of its deformation process, Metals and Materials 4 (1998) 911–914.
[25] Y. Nakiyama, M. Mochihara, M. Fujita, Dynamic punching of holes in an aluminum pipe using impulsive hydraulic pressure, Journal of Materials Processing Technology 85 (1–3) (1999) 204–208.
[26] M. Mochihara, M. Fujita, K. Hokamoto, J. Ueno, S. Itoh, Blanking process of a metal plate sheared by underwater shock wave [In Japanese.], Journal of the Japan Explosives Society 59 (6) (1998) 285–293.
[27] K. Shiramoto, A. Kobayashi, K. Takamiya, M. Fujita, An investigation on deburring technology using concentrated underwater shock wave, Materials Science Forum 767: (2013) 183–187.
[28] K. Shiramoto, J. Shimizu, A. Kobayashi, M. Fujita, A deburring process performed by underwater shock wave, Materials Science Forum 673 (2011) 271–274.
[29] S. Tanaka, I. Bataev, M. Nishi, I. Balagansky, K. Hokamoto, Micropunching large-area metal sheets using underwater shock wave: experimental study and numerical simulation, International Journal of Machine Tools and Manufacture 147 (2019) 103457.
[30] C. Zheng, X. Zhang, Y. Zhang, Z. Ji, Y. Luan, L. Song, Effects of laser power density and initial grain size in laser shock punching of pure copper foil, Optics and Lasers in Engineering 105 (2018) 35–42.

CHAPTER 3

Explosive welding

Akihisa Mori[a], Ivan Bataev[b], Naoyuki Wada[c], Kazuyuki Hokamoto[d]
[a]Sojo University, Kumamoto, Japan
[b]Novosibirsk State Technical University, Russia
[c]Asahi Kasei Corporation, Chikushino, Japan
[d]Professor, Institute of Industrial Nanomaterials, Kumamoto University, Kumamoto, Japan

3.1 Introduction

In the field of welding, the welding of dissimilar metals is a well-known challenge. In the use of welding through melting, it is well known that the welding of dissimilar metals is difficult. Solid-state welding, such as friction welding and friction stir welding (FSW), is the common candidate for such dissimilar metal joining, and FSW is currently used widely around the world for joining various metals, especially for light metals such as aluminum and magnesium alloys. FSW can provide dissimilar metal joints for various industrial demands, but the welding is possible only in line and the metal combinations are limited.

In contrast to regular welding techniques, explosive welding is unique because of the wider array of possible metal combinations [1]. In addition, it is possible to join the surfaces of plates over a large area. The processing time for the welding is extremely short, which may bring unique microstructural features at the bonding interface.

This chapter first presents the current status of explosive welding for industrial applications. After that, microstructural features at the welding interface are discussed, followed by the introduction of an improved method—underwater explosive welding, which is capable of welding various difficult-to-weld materials.

3.2 Basic issues and industrial application

Explosive welding is a technique to bond two or more, and similar or dissimilar, metals by means of an enormous energy burst released from a detonated explosive. The metal composites produced by this technique are called *explosion welding metal*, *explosion bonding metal*, and *explosive welding clad*, among others, and there are several companies manufacturing these composites around the world. These clads are used worldwide in areas such as chemical

Explosion, Shock-Wave and High-Strain-Rate Phenomena of Advanced Materials
DOI: 10.1016/C2019-0-03954-4

plants, oil refinery stations, petrochemical plants, and power plants. Typical applications of such clads are corrosion-resistant plates and sheets in condenser tube sheets of power plants, tube sheets, shells and head plates in heat exchangers of chemical plants, plate liners in rolling machines in steel mills, cryogenic transition joints, ultra-vacuum transition joints and electrical transition joints, and structural transition joints for ships and LNG carriers. Figs. 3.1 and 3.2 show examples of some industrial applications of these clads.

Fig. 3.1 ***Clad tube sheet used for a petrochemical heat exchanger.*** *(Photo provided by Asahi Kasei Corp).*

Fig. 3.2 ***Reactor for corrosion resistance using titanium clad.*** *(Photo provided by Asahi Kasei Corp).*

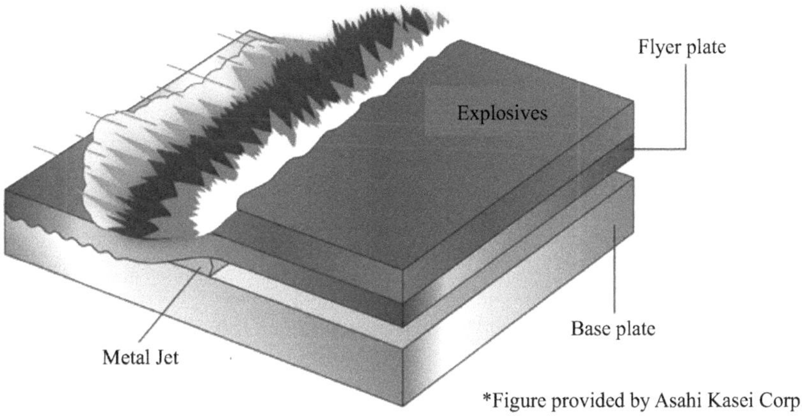

Fig. 3.3 ***Schematic illustration of the explosive welding process.*** *(Figure provided by Asahi Kasei Corp).*

The manufacturing process of explosive welding clads uses huge energy from explosions. It is possible to bring metallurgical bonding of more than two similar and/or dissimilar metals, and it is normally categorized as a cold welding process, but the real phenomena include quite an activated process called *metal jet* [1]. A schematic illustration of the welding process is shown in Fig. 3.3. Cladding metal (flyer plate) is positioned above the base metal (base plate) with a specific clearance between these plates. An explosive is charged over the surface of the flyer plate, and the explosive is detonated by an electric detonator where the detonation propagates with high speed at 2000 to 3000 m/s. Once the charged explosive is detonated, high pressure, in the order of several gigapascals, induced at the detonation front accelerates the flyer plate at several hundred meters per second and collides with the base plate at a certain inclination angle to induce the metal jet ahead of the collision point [1]. The metal jet enables the blowing up of surface oxides, adsorbed gas, and remaining contamination on the surface of the metals, and the cleared surfaces are pressurized higher than the flow strength of the metals, which may reach several tens of gigapascals [1]. Under such a situation, the activated metal surfaces are bonded tightly to each other by atomic attraction force. Explosive welding is completed within microseconds. The effect of heating is minor for such a short processing time, and it is well known that there is formation of a wavy interface, as seen in Fig. 3.4.

The moderating condition of explosive welding has been investigated by changing parameters such as collision velocity, collision angle, and

Fig. 3.4 ***Wavy interface found on an explosive welding clad (Nb/Cu).*** *(Photo provided by Asahi Kasei Corp).*

horizontal point collision velocity of the flyer plate. When the explosion is initiated, the flyer plate is accelerated at a high velocity V_p by detonation pressure and collides with the base plate at a collision angle β, as shown in Fig. 3.5. The flyer plate velocity V_p is related to the horizontal point collision velocity V_c and collision angle β, as shown in Eq. (3.1) [1]:

$$V_p = 2V_c \sin\left(\frac{\beta}{2}\right). \tag{3.1}$$

Collision angle β is almost proportional to flyer plate velocity V_p when the bending angle is small, and the angle β and horizontal collision point velocity V_c are the most important parameters to achieve successful welding. The condition is described as the explosive welding window, as shown in Fig. 3.6 [2,3]. The explosive welding window suggests the formations of the metal jet, the critical impact pressure, the maximum impact velocity and wavy–smooth transition velocity at the welded interface, and the details are discussed next.

The critical condition for the formation of the metal jet is shown as line (1) in Fig. 3.6. When the horizontal collision point velocity V_c exceeds

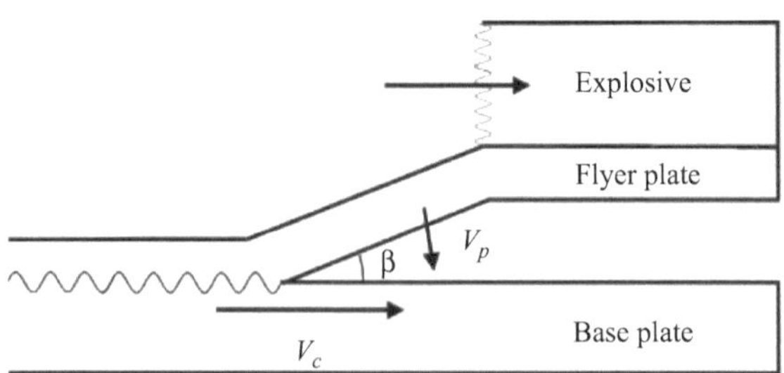

Fig. 3.5 ***Schematic illustration of explosive welding.***

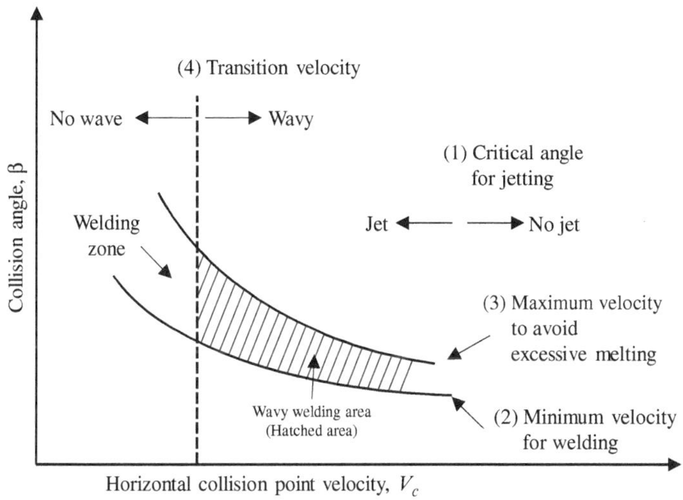

Fig. 3.6 ***Welding window.***

the sound velocity of metals, there is no jetting, and thus the jetting region, where V_c is lower than the sound velocity of metal, has the possibility of welding. The welding is not achieved when the impact pressure (impact velocity) is too low. Line (2), the minimum velocity for welding, shown in Fig. 3.6, is represented in Eq. (3.2) [4]:

$$\beta = k_1 \sqrt{\frac{H_v}{\rho V_c^2}}, \tag{3.2}$$

where β is the collision angle in radians, k_1 is a constant, H_v is the average Vickers hardness, and ρ is the density. The value of k_1 is 0.6 for high-quality precleaned surfaces and 1.2 for imperfectly cleaned surfaces.

The maximum velocity for the welding is also determined to avoid excessive melting, which induces brittle intermetallics at the welded interface. Line (3), the maximum velocity for welding, shown in Fig. 3.6, is represented in Eq. (3.3) [4]:

$$\sin\left(\frac{\beta}{2}\right) = \frac{k_2}{t^{0.25} V_c^{1.25}}, \tag{3.3}$$

where k_2 should be evaluated experimentally at a value of V_c that is equal to half of the compressive wave velocity and t is the thickness of the flyer plate.

The welding window is also capable of predicting the wave formation that would be able to determine the smooth–wavy transition based on experimental results, and wave formation can be described as being analogous to the fluid flow behind an obstacle. Thus, the smooth–wavy boundary corresponds to the "laminar-turbulent transition," so the Reynolds number characterizes this transition in fluid mechanics, and Reynolds number *Re* for explosive welding is introduced in Eq. (3.4) [5,6]:

$$Re = \frac{(\rho_F + \rho_B)V_t^2}{2(H_{VF} + H_{VB})} \tag{3.4}$$

where *F* and *B* are the densities of the metals; H_{VF} and H_{VB} are the Vickers hardness of the flyer and parent plates, respectively; and V_t is transition velocity. When Reynolds number *Re* used for Eq. (3.4) is set at 10.6, this equation can describe the transition appropriately [5,6]. Thus, line (4) suggests the transition velocity V_t, as shown in Fig. 3.6.

As described previously, the appropriate welding condition for each metal combination can be obtained, shown as the hatched area in Fig. 3.6.

Explosive welding clads have several unique product features:

1. Higher bonding strength than any other type of bonded products (roll bonding or overlaying).
2. Excellent in workability for bending and drawing.
3. A 100% bonding area leads to the following features:
   - No property degradation by the heat cycle.
   - Superior gas tightness (for transition joints).
   - No electric power loss at the bonding interface (for electroconductive applications).
4. Less alloy formation at the bonding interface and superior mechanical properties compared with other bonding products.
5. Almost all existing metals and alloys can be bonded. Especially titanium and aluminum, which cannot be overlaid, can be bonded, and tantalum and zirconium as well.
6. Forged steel can be bonded.
7. Manufacturing of multilayered clad plates is possible.
8. Cold process, free from the heat effect, so there is no deterioration of the original property of metals.

The mechanical properties of explosive welding clads such as shear strength, tensile strength, and ductility are better than in other bonding products. Shear strength is tested and guaranteed in accordance with ASTM, JIS, and so forth. Table 3.1 shows the shear strength of various clad plates. The tensile strength of the clad plates listed in the table also satisfies the values using Eq. (3.5), where δ_1 and δ_2 are the lower limit of the specified value for the tensile strength (MPa) of the flyer plate and base

Table 3.1 Shear strength of various clads.

Metal combination		Shear strength (MPa)	
Flyer plate	**Base plate**	**Actual data**	**JIS Standard**
Stainless steel	Carbon steel	300–400	Min. 200
Titanium	Carbon steel	200–350	Min. 140
Copper	Carbon steel	150–200	Min. 100
Aluminum bronze	Carbon steel	300–400	Min. 100
Naval brass	Carbon steel	200–250	Min. 100
Nickel alloy	Carbon steel	300–400	Min. 200
Nickel	Carbon steel	300–350	Min. 200
Monel	Carbon steel	300–400	Min. 200
Aluminum	Carbon steel	50–100	—

Note: Actual data is not a guaranteed value.

plate, and t_1 and t_2 are the thickness (mm) of the flyer plate and base plate, respectively:

$$\sigma = \frac{t_1\delta_1 + t_2\delta_2}{t_1 + t_2}. \tag{3.5}$$

Ductility of the flyer plate is guaranteed in the face bend (flyer plate is in tension), and the ductility of the base plate is guaranteed in the reverse bend (flyer plate is in compression). In addition, the lower limit of the elongation of the flyer plate or base plate is guaranteed.

As described at the beginning of this chapter, explosive welding clads are sometimes used after processing with bending, welding and so on. In case of usage for special circumstances, such as cryogenics or high vacuum, the transition joint manufactured by explosive welding is widely used. The combinations of such transition joints are copper/aluminum, stainless steel/aluminum, steel/aluminum, and stainless steel/copper.

The electrical transition joint composed of copper and aluminum and aluminum and steel, and so forth, have excellent electrical and mechanical properties, as well as good heat resistance and corrosion resistance. Thus, such joints are used for metal refining, salt electrolyzers, chemical plants, cryogenics, and heavy-duty electrical parts, among others.

The transition joint for cryogenic pipes is composed of aluminum and stainless steel, as illustrated in Fig. 3.7, and such joints are not allowed to have any leakage of gas or fluid. These joints need to maintain high strength under the cryogenic circumstance. Using the same transition joint, direct welding of stainless steel and aluminum, as shown in Fig. 3.8, the pipes are

Fig. 3.7 ***Transition joints for cryogenic parts.*** *(Photo provided by Asahi Kasei Corp).*

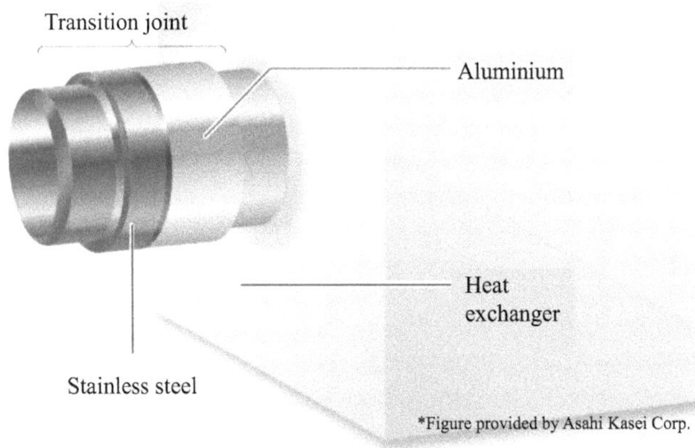

Fig. 3.8 ***Schematic illustration of the transition joint for a cryogenic pipe.*** *(Figure provided by Asahi Kasei Corp).*

not allowed to leak liquid or gas, which assures the advantage of environmental safety, and currently, small-diameter pipes are available for customers to provide compact pipeline designs. The bonding strength at the interface is stable under cryogenic circumstances, and it is also stable against heat cycles. The transition joint pipes for cryogenic temperatures are applied for air separation units, LNG·GTL (gas to liquid) plants, accelerators, and so forth.

The structural transition joint enables direct welding between aluminum alloy and carbon steel or stainless steel, as shown in Figs. 3.9 and 3.10. This joint is applied for ships using aluminum in the upper body and

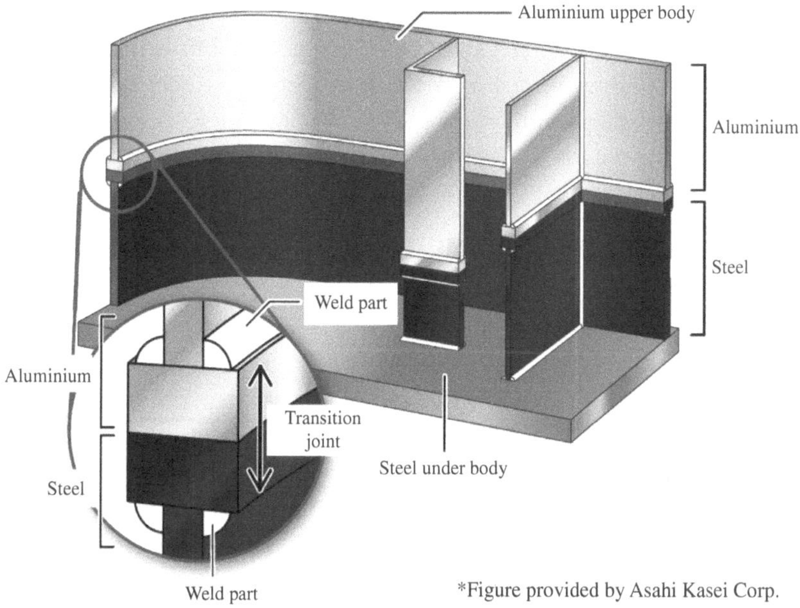

Fig. 3.9 ***Structural transition joint***. *(Figure provided by Asahi Kasei Corp).*

Fig. 3.10 ***Ocean cruiser using structural transition joint***. *(Photo provided by Asahi Kasei Corp. Courtesy of Mitsubishi Heavy Industries Ltd).*

carbon steel in the under body, and the use of such joints makes it possible to speed up ships, as well as brings energy savings based on corrosion resistance and water tightness. Structural transition joints have no risk of debonding due to the high bond strength at the interface.

In addition, other types of transition joints for transport equipment, such as airplanes, trains, and vehicles, are being investigated and developed. As an example, the use of noncombustible magnesium alloy clad is focused on decreasing the weight of transport equipment. This welding of magnesium with other metals is quite difficult due to the difference in the properties of metals, and magnesium easily induces brittle intermetallics through various conventional methods. Our recent research suggested that noncombustible magnesium alloy and aluminum can be bonded successfully with high strength using the explosive welding process [7].

3.3 Microstructures at the interface of explosively welded materials

The conditions of bond formation in the process of explosive welding and other types of high-velocity impact welding technologies significantly differ from those in various methods of fusion welding, diffusion welding, and even pressure welding. For instance, deformation and thermal processes during explosive welding most intensively occur in a narrow region adjacent to the interface between the plates, whereas the greatest part of the material remains practically undeformed and not subjected to the influence of high temperatures. Numerous studies indicate that the plastic strain exponentially increases in the direction of the interface. The results of multiple experimental observations are summarized in the monograph of Lysak and Kuzmin [8], whereas similar conclusions drawn based on numerical simulation can be found in the work of Émurlaeva et al. [9]. The temperatures, most probably, increase in the same way as the strains. Even though accurate data on the temperatures achieved during explosive welding have not been reported in the literature so far, it is nevertheless obvious that the temperature at the interface can significantly exceed the melting point of the welded alloys. The jet moving in front of the contact point can even be in a plasma state. The processes of plastic deformation, heating, and subsequent cooling occurring extremely quickly play an important role in the bond formation. The whole process of welding typically lasts from several microseconds to several milliseconds, depending on the bonding area. Thus, although the mixing zones during explosion welding are, as a rule, extremely small, they are essentially inhomogeneous both in chemical and phase compositions. Moreover, a high cooling rate due to the rapid heat transfer from the interface to unheated volumes of the welded plates often leads to quenching.

The combination of the preceding conditions leads to the formation of unique microstructures at the interface between the explosively welded workpieces, as well as at the adjacent areas. Such microstructures are rarely observed in welded joints obtained by other technologies. Due to the recent advances in materials characterization techniques, primarily due to progress in transmission electron microscopy (TEM), a significant amount of data on the features of microstructure formation at the interface has appeared in the literature in recent years. These data allow for a much better understanding of the bonding phenomena and may be used to explain some properties of explosively welded materials. In this section, we summarize the current understanding of such structures and the reasons behind their formation.

In discussing the microstructures at the interface, it makes sense to consider separately the structural transformations occurring in the mixing zones and in the zones located at some distance from them.

Materials intermixing during explosive welding is a well-known fact. Microscopic and elemental analysis of welded joints often reveal microvolumes, which consist of elements of both welded plates. Depending on the collision regimes, these mixing zones can have different shapes. For example, they can be found along the entire surface of welded plates in the form of an almost continuous film with a thickness ranging from fractions of a micrometer to several tens of micrometers (Fig. 3.11A) [10]. Under collision regimes leading to wave formation, mixing zones are typically formed as discrete microvolumes located on the sides of the waves (Fig. 3.11B) [10]. In this case, the mixing zones are often called *vortices* due to the peculiarities of the flow twisting in these microvolumes. In addition, between the two cases described earlier, a whole spectrum of morphologies of mixing zones has been observed in different experimental studies.

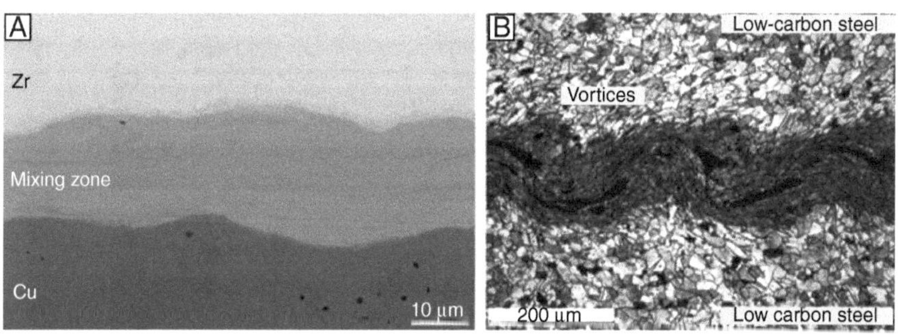

Fig. 3.11 ***(A) A continuous mixing layer along a flat interface formed by explosive welding of Zr to Cu. (B) A wavy interface with discrete vortex zones formed by explosive welding of two thin plates of low-carbon steel.*** *(From I. A. Bataev, D. V. Lazurenko, S. Tanaka, K. Hokamoto, A. A. Bataev, Y. Guo, et al. High cooling rates and metastable phases at the interfaces of explosively welded materials.* Acta Materialia *135 (2017) 277–289).*

In the scientific community, there is ongoing discussion regarding the state of material in the mixing zones during the process of explosive welding. There are different opinions as to whether the melting occurs at the interface and, if so, what the role of melting is in bond formation. A significant number of metallographic studies indicate that mixing at the interface occurs in a liquid state. This is primarily evidenced by the structures and morphologies typical of primary crystallization. In particular, one can frequently find a dendritic structure of the mixing zones by fractographic analysis of welded joints. In such a case, thin dendrites consisting of main axes and poorly developed secondary axes growing against the direction of the heat transfer are observed in the mixing zones [11-14]. However, when the welding regimes are selected closer to the lower welding limit, the melting zone becomes very thin and sometimes it is not observed using light microscopy or scanning electron microscopy.

A study of Hammerschmidt and Kreye [15] was one of the first in which an assumption was made that the formation of bonding during explosive welding occurs due to the local melting at the interface and rapid solidification of the molten layer. During a TEM study of explosively welded plates of Al-Cu alloy, they observed a thin interlayer consisting of equiaxed grains less than 300 nm in size formed at the interface. The most logical reason for the formation of this layer proposed by Hammerschmidt and Kreye was the local melting of the material and subsequent rapid solidification.

In the middle of the mixing layers, cavities are frequently observed (see some examples in other studies [13,16-18]). It should be noted that the size of these cavities is typically too large to explain them solely as a result of solidification shrinkage. In addition, the formation of such cavities has been observed in recent studies devoted to smooth particle hydrodynamics simulation of high-velocity impact welding (i.e., [9]).

Next, we consider the experimental data on the structure and phase composition of the mixing zones formed during explosive welding of various materials. For ease of analysis, all materials were classified into the following groups: (1) the welded plates have the same elemental composition; (2) the welded plates consist of materials that are completely soluble in each other in liquid and solid state; (3) the welded plates consist of immiscible or poorly miscible alloys; and (4) the welded plates consist of alloys, the mixing of which leads to the formation of intermetallic phases.

First, explosive welding of materials with the same composition is rarely carried out in practice; nevertheless, it may be of interest from a fundamental point of view. For example, a significant amount of data clarifying the

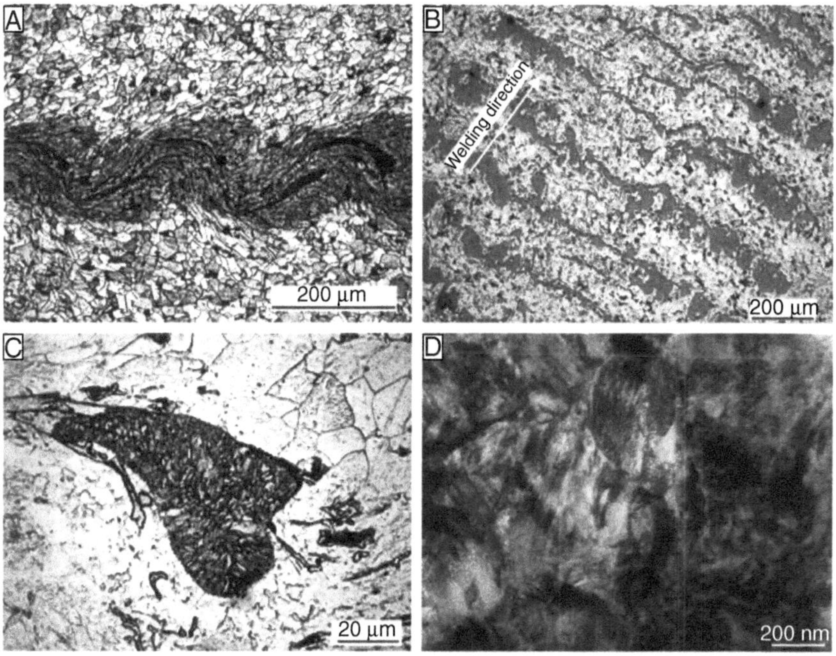

Fig. 3.12 ***(A, B) The interface of explosively welded plates of low-carbon steel studied in two mutually perpendicular sections. (C, D) Enlarged images of the vortex zone captured by light microscopy and scanning electron microscopy, respectively.*** *(Images provided by I. A. Bataev).*

features of the joint formation during explosive welding can be obtained by analyzing steel/steel joints.

In the mixing zones of explosively welded low-carbon steel plates, one almost always observes a martensitic structure regardless of the welding conditions (Fig. 3.12) [10,19,20]. Moreover, due to the high cooling rate, the hardness of the martensite is quite high even if the content of carbon was quite low. For instance, it was more than 500 HV for plain carbon steel with 0.2 wt.% C [10]. It should be noted that the critical cooling rate required for quenching of low-carbon steels should be quite high (for steel with 0.2% of carbon, it should exceed 1000 K/s [21]). It is also interesting to note that the wavy shape of the interfaces observed in the cross section of the explosively welded samples actually has a complex spatial configuration. This means that some instability of plastic flow occurs not only in the welding direction but also in the transverse direction. This becomes apparent when studying the shape of the interface in two mutually perpendicular sections (Fig. 3.12A and B).

Explosively welded multilayer composite consisting of commercially pure titanium was studied by Makarova et al. [22]. The mixing zones in this material had a needle-like martensitic structure typical of quenching of titanium alloys. The mixing zones had a fairly high hardness (more than 800 HV) despite the absence of a significant number of alloying elements, which was probably caused by the dissolution of oxygen in the molten material.

Second, explosive welding of mutually soluble materials will be considered using the example of a copper/brass joint. This combination was described by Bataev et al. [10]. Mixing zones in explosively welded pure copper and copper-zinc brass (63 wt. % Cu, 37 wt.% Zn) consisted only of α-Cu. On average, the mixing zones consisted of 20 wt.% Zn and 80 wt.% Cu. The formation of only an α-Cu phase in an alloy of such a composition seems to be quite reasonable considering the fact that no data have been reported on phase transformation in this alloy induced by quenching or application of high pressures. It is interesting to note that the formation of nanoscale pores was observed using TEM. Considering the low boiling point of brass containing a large amount of highly volatile zinc, the possibility of momentary boiling of material in the vortex zones during explosive welding cannot be ruled out. It should be noted, however, that the fact of materials boiling during explosive welding has not been discussed in the literature so far and requires thorough experimental verification. Nevertheless, in a significant number of studies, small pores of spherical shape have been observed in the mixing zones [10,23].

Third, explosively welded bimetals consisting of materials that are immiscible in solid state include, but are not limited to, the following combinations: Ta/Cu, Fe/Ag, Fe/Mg, Ti/Mg, W/Cu.

Fig. 3.13 shows the structures typical for explosively welded tantalum and copper, which are almost immiscible in the solid state [10,24]. In the mixing zones, tantalum forms either micro- or nanoscale nearly spherical particles located in a copper matrix (Fig. 3.13A [10], Fig. 3.13B [24]), or a peculiar sponge-like structure, the cavities of which are filled with copper (Fig. 3.13C [24], Fig. 3.13D [24]). In a study by Greenberg et al. [25], the phrase "solidified colloidal solution of two insoluble phases" was used to describe a structure consisting of nano-sized tantalum particles in a copper matrix. According to Ta-Cu equilibrium phase diagram, this system is miscible above the liquidus; however, between the liquidus and solidus, there is a miscibility gap leading to the separation of two liquids into tantalum-rich and copper-rich phases. It can be assumed that the nanoscale

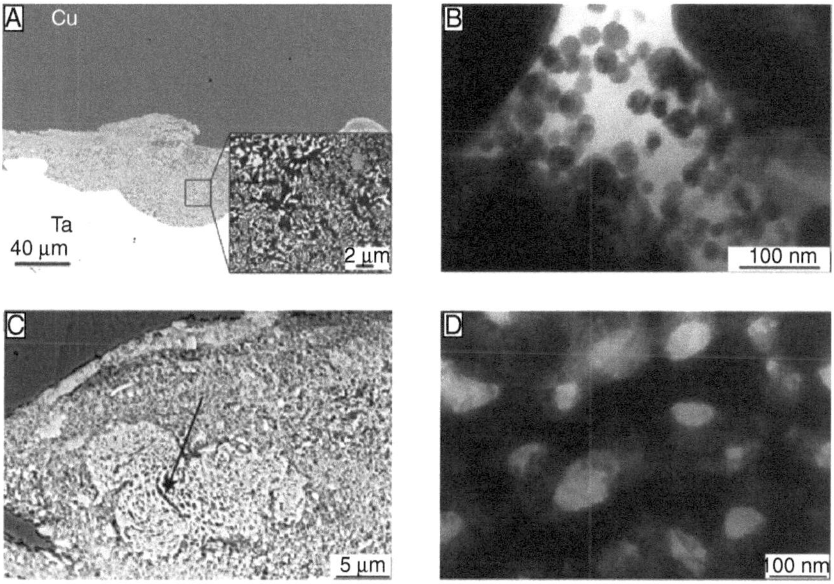

Fig. 3.13 ***The structures near the interface of explosively welded Cu/Ta bimetal.*** (A, B) Micro- and nanocrystalline tantalum particles distributed in a copper matrix. (C, D) Tantalum "sponge" filled with copper. *(Part (A) from I. A. Bataev, D. V. Lazurenko, S. Tanaka, K. Hokamoto, A. A. Bataev, Y. Guo, et al. High cooling rates and metastable phases at the interfaces of explosively welded materials.* Acta Materialia *135 (2017) 277–289. Parts (B) through (D) from I. N. Maliutina, V. I. Mali, I. A. Bataev, A. A. Bataev, M. A. Esikov, A. I. Smirnov, et al. Structure and microhardness of Cu-Ta joints produced by explosive welding.* Scientific World Journal *2013 (5) (2013) 256758).*

colloidal solution was formed between the solidus and liquidus as a result of phase separation and the pulsed impact on the mixing zone. Due to the high cooling rate, the coagulation process was not completed, and the system quickly solidified, which explains the small size of the Ta particles. It is likely that nanoscale spherical tantalum particles in a copper matrix observed in Fig. 3.13B were formed from liquid enriched with copper, and the spongy structure of the tantalum appeared in areas enriched with tantalum (Fig. 3.13D). TEM analysis performed by Maliutina et al. [24] did not reveal any phases in mixing zones other than tantalum and copper. It is interesting to note that in a recent study Parchuri et al. [26] showed that in explosively welded Ta/Cu bimetal, along with tantalum and copper phases, the formation of a metastable quasicrystalline phase is possible.

The structures observed between other immiscible materials welded by explosion were similar to those observed in Cu/Ta joints. For instance, in the case of welding steel with silver, the mixing zones consisted of iron particles in a silver matrix [27]. When welding copper and tungsten, mixing zones consisting of subnanometer tungsten particles distributed in a copper matrix were observed [28].

Fourth, the analysis of phase diagrams indicates that most binary metallic systems tend to form various intermetallic compounds. With rare exceptions, intermetallic compounds formed between metals are very brittle. For this reason, joining dissimilar metals using methods based on fusion welding is an extremely difficult task. The joints formed using such an approach contain large volumes of intermetallic phases, which are prone to brittle fracture. Thus, fusion-welded dissimilar joints frequently cannot withstand the thermal stresses induced at the cooling stage.

Among explosively welded bimetals, most systems are also prone to the formation of intermetallic compounds. Such bimetals include, for example, titanium/steel, aluminum/copper, aluminum/iron, and many other combinations. However, in comparison with the joints obtained by fusion welding, the mixing zones in explosively welded joints are significantly smaller (from several tens of nanometers to several tens of micrometers). This fact probably explains the significantly better quality of explosively welded bimetals in comparison with bimetals obtained by other joining techniques. In addition, there is a point of view according to which, with correctly selected welding conditions, it is possible to completely prevent melting at the interface and pro-vide a direct bonding between two materials in the solid phase.

Many intermetallic compounds have a complex crystal lattice. The arrangement of atoms in a thermodynamically stable long-range ordered structure during primary crystallization requires significant displacements of individual atoms by a diffusion mechanism. For this reason, rapid solidification of molten intermetallic alloys frequently leads to the formation of metastable phases. For instance, many metallic glasses and quasicrystalline alloys are formed by rapid solidification of alloys, the slow cooling of which leads to formation of crystalline intermetallics.

One of the first studies showing formation of metastable phases in the mixing zones was by Nishida et al. [29,30]. Using TEM, they observed amorphous phases at the interface of explosively welded steel and titanium composite and icosahedral quasicrystals, as well as amorphous phases at the interface of explosively welded nickel and titanium composite. In addition, in Ni/Ti bimetal, they observed formation of NiTi and Ti_2Ni crystalline phases. The formation of metastable phases was explained by local melting

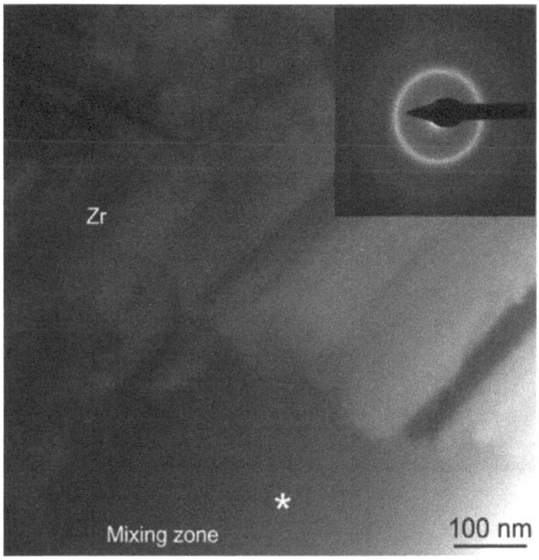

Fig. 3.14 ***Mixing zone formed during explosive welding of zirconium and copper, consisting mainly of an amorphous phase (see the electron diffraction pattern in the inset).*** *(From I. A. Bataev, D. V. Lazurenko, S. Tanaka, K. Hokamoto, A. A. Bataev, Y. Guo, et al. High cooling rates and metastable phases at the interfaces of explosively welded materials.* Acta Materialia *135 (2017) 277–289).*

followed by subsequent rapid solidification of the molten microvolumes. The formation of quasicrystals during explosive welding of the Ni/Ti composite was recently confirmed in a study by Bataev et al. [31]. In another study by Bataev et al. [10], it was reported that in the mixing zones of a low-carbon steel and titanium alloy composite, the formation of an amorphous phase, icosahedral quasicrystals, and nano-sized crystals of $TiFe$ and $TiFe_2$, was observed. Similar phases were found in a study by Lazurenko et al. [32].

The mixing zones formed in explosively welded Zr and Cu sheets possessed a predominantly amorphous structure (Fig. 3.14 [10]). Using energy-dispersive X-ray spectroscopy, it was found that the mixing zones consisted mainly of copper (65–70 wt.%). It is interesting to note that such composition is typical of binary metallic glasses of the Cu-Zr system. From the literature, it is known that the alloys of this system are one of the best glass formers among binary metallic alloys. Moreover, some alloys of the Cu-Zr system are able to form bulk metallic glasses even without additional alloying. Considering the high cooling rates typical of explosive welding [9,10,31], the formation of an amorphous structure in this composite was quite expected.

A small amount of amorphous phase was also found in Ni/Zr composite; however, along with it, a significant amount of crystalline intermetallics was also detected [10]. For instance, $Ni_{10}Zr_7$, $Ni_{21}Zr_8$, and $ZrNi_5$ crystalline precipitations were observed. The formation of a higher number of crystalline phases in the Ni/Zr composite compared with the Cu/Zr composite was for two reasons. First, nickel has a lower thermal diffusivity compared with copper, so it can be expected that under otherwise equal conditions, the cooling rates during welding of this composite should be lower. Second, Ni-Zr alloys have a lower glass-forming ability in comparison with Cu-Zr alloys.

At the interface of explosively welded pure aluminum and AZ31 magnesium alloy, the formation of two phases was observed (Fig. 3.15 [10]). One of them—$Al_{12}Mg_{17}$ (Fig. 3.15C)—could be found on the equilibrium phase diagram of the Al-Mg system. The second phase was metastable and was the so-called incommensurate phase with a modulated structure (Fig. 3.15D). This phase was first discovered in the Al-Mg system by Kobayashi et al. [33] during melt spinning of an Al alloy 34.3 at.% Mg. The cooling rate achieved in the experiments of Kobayashi et al. [33] was 104 to 105 K/s. Kulkarni et al. [34] called this compound the *pseudo icosahedral phase*. Donnadieu et al. [35] and Elcoro and Perez-Mato [36] claimed that this phase possesses a cubic incommensurate modulated structure and called it a *cubic quasicrystal*.

In the work of Bataev et al. [31], the formation of a mixture of amorphous and quasicrystalline phases was observed in explosively welded bimetals Ta/stainless steel and aluminum/niobium.

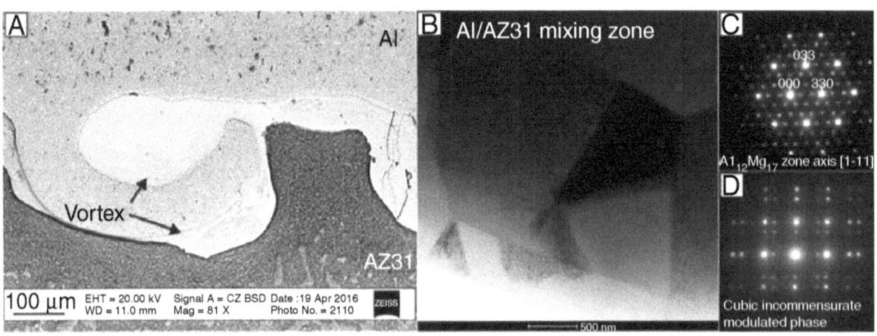

Fig. 3.15 ***Explosively welded Al/AZ31 composite.*** (A) Scanning electron microscopy image of vortex zones formed at the interface, (B) transmission electron microscopy image of the vortex zone microstructure, and (C, D) electron diffraction patterns of Al12Mg17 and cubic incommensurate modulated structures, respectively. *(From I. A. Bataev, D. V. Lazurenko, S. Tanaka, K. Hokamoto, A. A. Bataev, Y. Guo, et al. High cooling rates and metastable phases at the interfaces of explosively welded materials.* Acta Materialia *135 (2017) 277–289).*

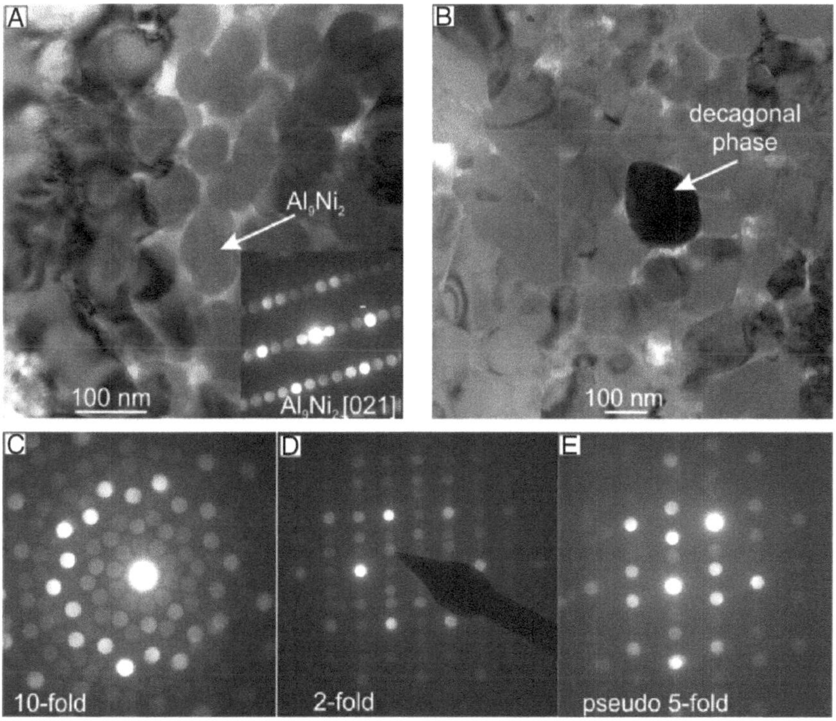

Fig. 3.16 ***Transmission electron microscopy images of the mixing zone in explosively welded nickel and aluminum composites showing precipitations of the Ni2Al9 metastable crystalline phase (A) and decagonal quasicrystalline phase (B–E).*** *(From I. A. Bataev, T. S. Ogneva, A. A. Bataev, V. I. Mali, M. A. Esikov, D. V. Lazurenko, et al. Explosively welded multilayer Ni-Al composites.* Materials and Design *88 (2015) 1082–1087).*

In the mixing zones of aluminum and nickel multilayer composite, a complex mixture of phases was observed [12]. In particular, nano-sized crystals of the AlNi, Al_3Ni, and Al_9Ni_2 phases, as well as the decagonal quasicrystalline phase, were detected using TEM investigation (Fig. 3.16 [12]). It should be noted that the Al_9Ni_2 phase and the decagonal phase in Al-Ni alloys are metastable and were previously discovered in rapidly solidified ribbons obtained by the two-piston splat cooling method.

A significant number of metastable phases was observed by Lazurenko et al. [37] in explosively welded commercially pure titanium and aluminum alloy Al-1 wt.% Mn [37]. For instance, the formation of a disordered $AlTi_3$ phase was observed. This phase, when obtained by slow cooling, forms as an ordered solution of aluminum in the titanium lattice. In this case, the solid solution disordering could have occurred as a result of rapid cooling. The most unexpected fact was the formation of amorphous phase in the mixing zones of this composite. The binary Ti-Al system does not correspond to the well-known empirical criteria for glass formation, and information on the formation of amorphous phases in this system by rapid solidification of melt is not currently available in the literature. Thus, the formation of amorphous phases in this system indicates that the unique conditions of solidification arise at the interface of explosively welded materials.

Paul et al. [38] carried out detailed TEM examinations of carbon steel/zirconium, stainless steel/zirconium, carbon steel/titanium and stainless steel/tantalum explosively welded joints. In all of the samples, the mixing zones contained partially amorphous regions, as well as nano-sized crystalline precipitates. Based on the elemental composition analysis, they concluded that the crystalline phases are often metastable since they cannot be found on the corresponding phase diagrams. In addition, they claimed that the metallurgical bond was achieved through local melting of the material at the interlayer boundaries since all of the studied joints contained thin layers of melt. In another study by Paul et al. [39], it was shown that the thickness of the molten interlayers between explosively welded stainless steel and tantalum plates could be as small as 15 to 30 nm; however, the melting always occurred when a high-quality joint was formed.

In a recent study, Paul et al. [40] analyzed an explosively welded Ti/Cu composite. They claimed again that the formation of a molten layer occurred along the entire interface. Even the areas located at a considerable distance from the large vortex zones contained thin interlayers (thickness of 50–200 nm) of molten metal. They concluded that the formation of a high-quality joint by explosive welding is always associated with local melting. Thus, the "solid phase/solid phase" boundaries observed with light microscopes are in fact more complex boundaries of the "solid phase/solidified liquid phase/solidified liquid phase/solid phase" type. Moreover, due to the high glass-forming ability of Ti-Cu alloys, the formation of a partially amorphous structure in the molten layer was observed. The formation of an amorphous phase in an explosively welded Ti/Cu composite was also reported earlier in a study by Zu et al. [41]. In the study on Al/Cu composites, Paul et al. [11] found several phases that are not observed on the equilibrium state diagrams of the Al-Cu system.

It may be noticed that amorphous structures in the vortexes can be frequently predicted using the empirical criteria of glass formation [42]. This analysis was successfully applied in the work of Bataev et al. [43] to explain formation of metallic glasses in explosively welded Nb/SUS 304 bimetal.

Pressure is another important factor that should be considered when the formation of mixing zone structure is discussed. The solidification of the material at the interface occurs during the action of compressive stresses. Pressure has a contradictory effect on the formation of metallic glasses during the rapid solidification. For example, from thermodynamics, it is known that under the action of compressive stresses, as a rule, the phase with the smallest volume is stabilized [44]. In the majority of cases, the crystalline phase has a lower specific volume compared with the amorphous phase with the same chemical composition. This is because the ordered packing of atoms makes it possible to arrange them more compactly than chaotically. At the same time, several experimental studies showed that the tendency to vitrification increased significantly if high pressures were applied to the melt during solidification [45–47]. The positive effect of pressure during rapid solidification is explained by an increase in the viscosity of the melt, which reduces the mobility of atoms.

A further part of this section will be devoted to a brief discussion of the structure of the material located near the interface, in which temperature was below the melting point. However, it will be considered in less detail.

Material located near the mixing zone undergoes significant shear deformation and concomitant heating. However, the heating temperature is not sufficient for its melting, so the formation of the microstructure occurs in the solid state and can be described in accordance with existing ideas about the thermomechanical processing. In addition to the temperature and deformation effects, to understand the peculiarities of the forming microstructures, it is also necessary to keep in mind that deformation occurs at very high rates and at extremely high pressures.

When discussing the structure of a material that has not been melted, one should consider the fact that the shear strain in explosively welded materials increases almost exponentially approaching the interface [8]. In addition, the thickness of the strongly deformed layer is rather small in comparison with the total thickness of the plates being welded; obviously, it depends on the welding conditions, but as a rule, it is less than 10% of the total thickness of the plates. Furthermore, due to the high strain rate, the generated heat has a significant effect on the structure formation. As a result, the structure of the material near the interface is determined by the competing processes of strain hardening and thermal softening. Thus, depending on the alloys and welding conditions, all structures typical of

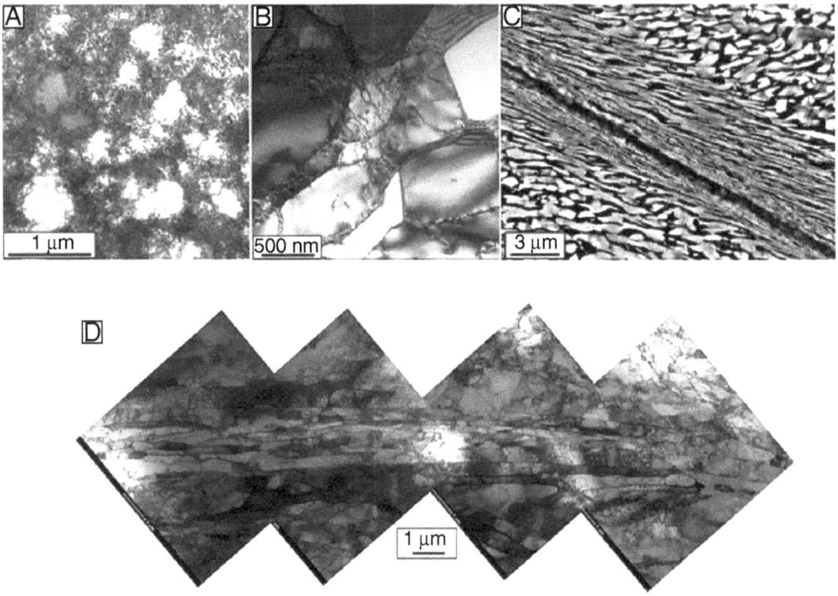

Fig. 3.17 ***Material structures formed in explosively welded materials in the solid state.*** (A) The cellular structure of nickel in the Ni/Al joint. (B) The polygonal structure of aluminum in the Ni/Al joint. (C) The adiabatic shear band in explosively welded plates of $\alpha + \beta$ titanium alloy. (D) The microband structure of ferrite in the joint of carbon steel to austenitic stainless steel. *(Parts (A) and (B) from I. A. Bataev, T. S. Ogneva, A. A. Bataev, V. I. Mali, M. A. Esikov, D. V. Lazurenko, et al. Explosively welded multilayer Ni-Al composites.* Materials and Design *88 (2015) 1082–1087).*

materials deformed at extremely high strain rates and deformation degrees are observed near the interface. Namely, after explosive welding, one may frequently meet dislocation cells (Fig. 3.17A) [12], polygonal structure (Fig. 3.17B) [12], adiabatic shear bands (Fig. 3.17C), microband structure (Fig. 3.17D), and other structures.

The effect of high pressures on the deformation process during explosive welding is rarely discussed in the literature. First, high compressive stresses provide the deformation with very high strain without failure. Second, in some cases, high pressures in combination with high temperatures can lead to a variety of phase transitions. For example, during high-velocity impact welding of steel plates, one may observe solid state $\alpha \rightarrow \gamma \rightarrow \alpha$ and $\alpha \rightarrow \varepsilon \rightarrow \alpha$ transformations near the interface [48]. In addition, the pressure may significantly increase the melting point and affect the volume fraction of molten zones in explosively welded materials.

3.4 Underwater explosive welding

Explosive welding is well known as one of the welding techniques applicable for joining dissimilar metals and has been widely industrialized because of its high bonding strength [1,49]. However, explosive welding is sometimes difficult, such as in cases that (1) use a thin flyer plate, (2) brittle materials, and (3) a combination of metals leading to excessive melting or reaction. To improve the welding condition, experimental parameters should be selected carefully or the parameters changed drastically. For this purpose, a new explosive welding assembly using an underwater shock wave was proposed and developed by Hokamoto and others, as schematically shown in Fig. 3.18 [50–51]. The details of the welding process are illustrated in Fig. 3.19.

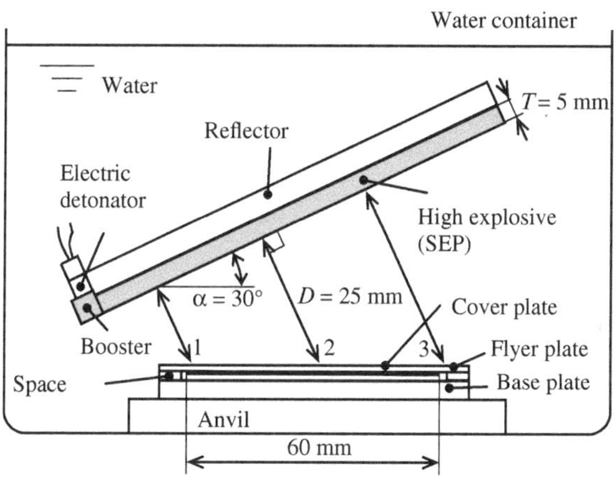

Fig. 3.18 ***Schematic of an underwater explosive welding technique.***

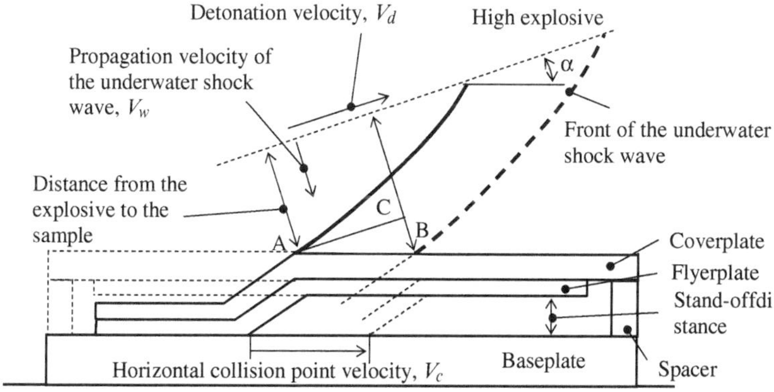

Fig. 3.19 ***Behavior of an underwater shock wave and welding parameters.***

In Fig. 3.18, the explosive used was SEP (detonation velocity 7 km/s, density 1310 kg/m^3) mainly composed of PETN and wax, and the explosive was set at a certain inclination angle α. To satisfy the condition of explosive welding, it is necessary to decrease the horizontal collision point velocity V_c to less than the sound velocity of materials [50,52], and such an inclined setup was chosen. The use of an underwater shock wave enables acceleration of the thin flyer and cover plates quite uniformly, so the welding of thin plates is possible, keeping the upper surface of the plate to be welded in the same condition as before the welding. Using the shock equations including the Hugoniot relation [53], the underwater shock pressure was estimated as 0.5 to 2 GPa depending on the horizontal position as an example. The pressure is high where the position is close to the detonator because of the small water distance from the explosive. The pressure is decreased with horizontal distance from the detonator side (left to right in Fig. 3.18).

Based on Fig. 3.19, the horizontal collision point velocity V_c, when the underwater shock wave moves from point A to B on the sample surface, can be calculated using Eq. (3.6):

$$V_c = \frac{AB}{\Delta t} = \frac{AB}{(AC/V_d) + (BC/V_w)} = \frac{AB}{(AB\cos\alpha/V_d) + (AB\sin\alpha/V_w)} = \frac{1}{(\cos\alpha/V_d) + (\sin\alpha/V_w)}. \tag{3.6}$$

The change in the horizontal collision point velocity V_c with the setup angle α for an explosive SEP (V_d = 7 km/s) is illustrated in Fig. 3.20 under the same experimental assembly as shown in Fig. 3.18. The value V_c is required to be less than the sound velocity of the metals to be welded [50,52]; a setup angle α more than 10 degrees is required, although the condition is dependent on the metals to be welded; and the pressuring conditions, shown as positions 1 (left edge of flyer plate), 2 (middle), and 3 (right edge) in Fig. 3.18, although the change of V_c is minor for the position.

In underwater explosions, it is known that intense repetitive pressurizing phenomena caused by the expansion and contraction of detonation gas, called the *bubbled pulse*, is induced when the system is closed [54]. For the present case, the effect of bubbled pulse is not considered since the flyer plate is accelerated only by the first underwater shock wave within the short stand-off distance. The detonation gas is easily released outside due to the small size of the water container.

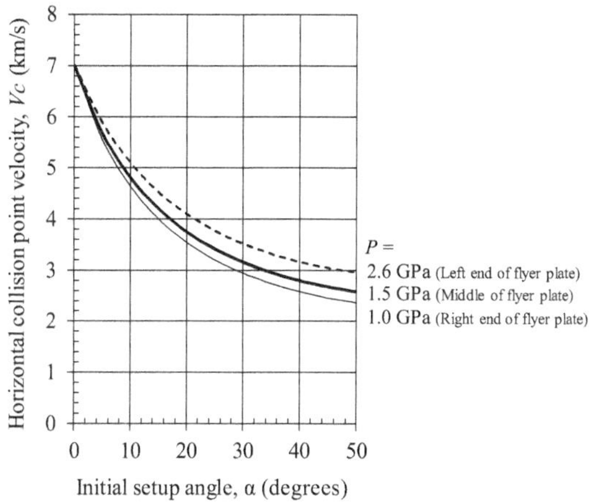

Fig. 3.20 ***Relationship between the collision point velocity and the initial inclined angle of explosive.***

One of the important advantages of underwater explosive welding is that it enables the welding of thin, difficult-to-weld plates. Some attempts have been made for welding an amorphous foil to another metal plate using the conventional explosive welding method [55-57], but the control of the welding parameters seems to be limited and using underwater shock waves makes it much easier to control the welding condition just by changing the dimensions of the assembly. Figs. 3.21 and 3.22 are the upper surface

Fig. 3.21 ***Top view of the recovered sample in the amorphous film (thickness of 38 μm) and stainless steel plate.*** *(From K. Hokamoto, K. Nakata, A. Mori, S. Tsuda, T. Tsumura, and A. Inoue. Dissimilar material welding of rapidly solidified foil and stainless steel plate using underwater explosive welding technique.* Journal of Alloys and Compounds *472 (1-2) (2009) 507–511).*

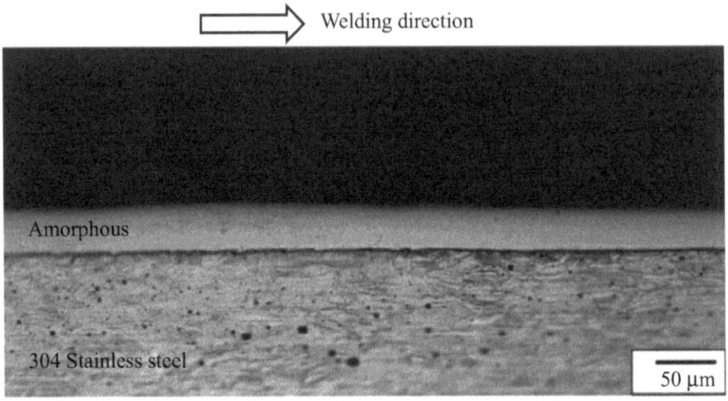

Fig. 3.22 ***Microstructure of longitudinal cross section along the welding direction on the welded amorphous film.*** *(From K. Hokamoto, K. Nakata, A. Mori, S. Tsuda, T. Tsumura, and A. Inoue. Dissimilar material welding of rapidly solidified foil and stainless steel plate using underwater explosive welding technique.* Journal of Alloys and Compounds *472 (1-2) (2009) 507–511).*

and the interface of an amorphous film welded on a 304 stainless steel plate [58]. The use of a thin flyer plate brings other advantages to this method. The use of a small-mass flying plate, normally flyer and cover plates, enables a decrease in the kinetic energy dissipated by the collision. The decrease in the kinetic energy lost by the collision may reduce the separation caused by excessive melting or reaction between the metals to be welded, which is related to the upper limit of welding [1,49].

As mentioned previously, the pressurizing condition should be changed with the horizontal position when using a setup with a fixed-thickness explosive, as shown in Fig. 3.19. Higher pressure is expected close to the detonator side and is decreased with increasing water distance in the horizontal direction. To prevent the reduction of the pressure with position, the use of an explosive with linearly increased thickness was proposed, as shown in Fig. 3.23 [26,58-61].

In underwater explosions, the peak pressure of the underwater shock wave at a distance from the explosive to sample D is given by Eq. (3.7) [54]:

$$P_{\max} = K\left(\frac{W^{1/\varepsilon}}{D}\right)^{n}, \tag{3.7}$$

where P_{max} is the peak pressure, K and n are material constants depending on the type of explosive, W is the mass of the explosive, and ε is the dimension where the explosive exists (i.e., $\varepsilon = 3$ when the phenomenon is a 3D problem). When the width of the explosive is sufficiently large

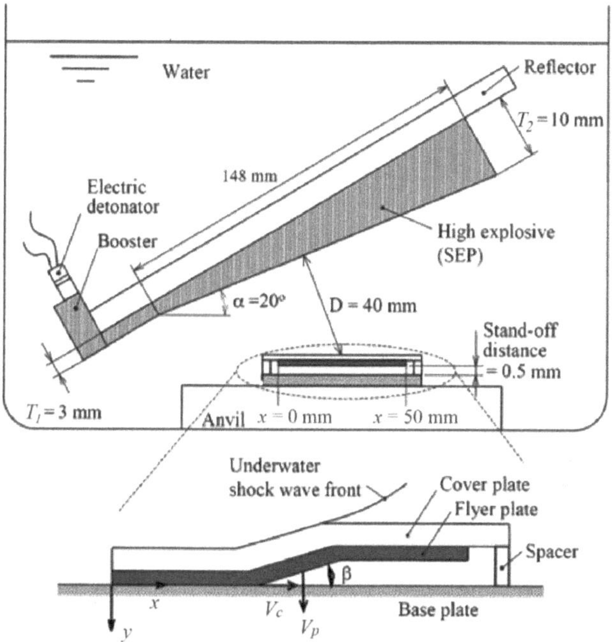

Fig. 3.23 ***Experimental setup for underwater explosive welding when an explosive with linearly increased thickness is used.*** *(From K. Hokamoto, K. Nakata, A. Mori, S. Tsuda, T. Tsumura, and A. Inoue. Dissimilar material welding of rapidly solidified foil and stainless steel plate using underwater explosive welding technique.* Journal of Alloys and Compounds *472 (1-2) (2009) 507–511).*

compared with the size of the sample, the effects on the length and width of the explosive can be ignored. Therefore, the behavior of the underwater shock wave can be considered as a 1D problem ($\varepsilon = 1$). In addition, it is possible that the mass of explosive W can be replaced with the thickness of explosive T, and Eq. (3.7) can be expressed as shown in Eq. (3.8):

$$P_{\max} = K\left(\frac{T}{D}\right)^{n}. \tag{3.8}$$

From the equation, it is possible to fix the peak pressure only by taking a uniform T/D ratio. Thus, increasing the thickness of explosive by increasing water distance is essential to keep the pressure uniform through the welding length. Based on the measurements of the size of interfacial waves, the wavelengths observed were almost uniform through the welding length using the modified assembly [59].

It is difficult to observe the high-velocity collision or welding phenomena through optical observation because of the narrow gap between the

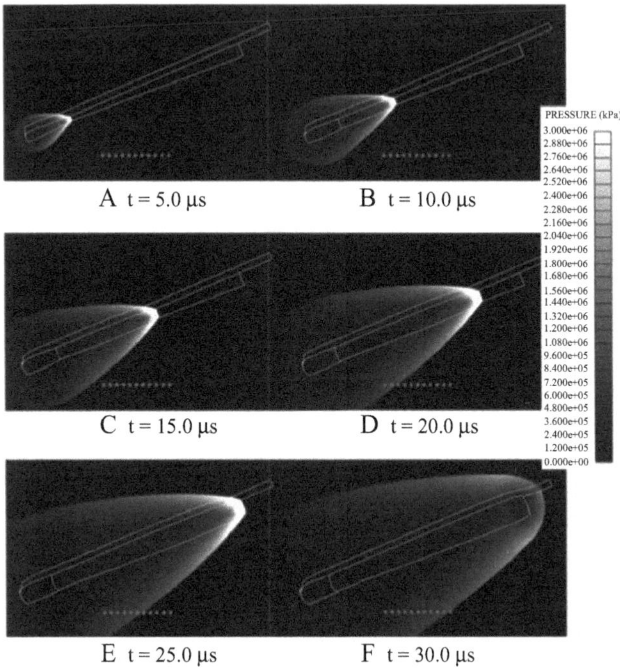

Fig. 3.24 ***Pressure contour of the underwater shock wave simulated by ANSYS AUTODYN-2D.*** *(From K. Hokamoto, K. Nakata, A. Mori, S. Tsuda, T. Tsumura, and A. Inoue. Dissimilar material welding of rapidly solidified foil and stainless steel plate using underwater explosive welding technique.* Journal of Alloys and Compounds *472 (1-2) (2009) 507–511).*

plates. Therefore, numerical analysis was made for the acceleration of the flying plate through the propagation of the underwater shock wave driven by detonation of the explosive. The analysis was made under the condition that the width is large enough, so the phenomena can be considered as a 2D problem and a 2D simulation was made to analyze the condition of explosive welding. As an example, the numerical results using ANSYS AUTODYN-2D are demonstrated in Fig. 3.24 [58]. The details of the parameters used are found in the work by Century Dynamics [62]. For the numerical simulation, explosive, reflector, and water parts were modeled by Euler solver, and the flyer and cover plates were applied by Shell solver, which is suitable for simulating the deformation of thin structures. The base plate was not modeled because it is only necessary to know the collision velocity V_p to the vertical direction or collision angle β and horizontal collision point velocity V_c to draw the welding window [2] so as to discuss the welding condition, weldable or not. Fig. 3.25 [58] summarizes the

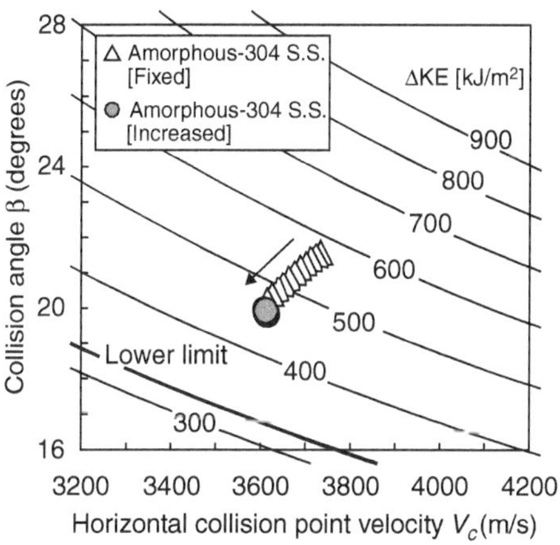

Fig. 3.25 ***Welding window comparison under the conditions using linearly increased thickness explosive (circles) and using fixed-thickness explosive (triangles).*** *(From K. Hokamoto, K. Nakata, A. Mori, S. Tsuda, T. Tsumura, and A. Inoue. Dissimilar material welding of rapidly solidified foil and stainless steel plate using underwater explosive welding technique.* Journal of Alloys and Compounds *472 (1-2) (2009) 507–511).*

numerical results as the welding window showing the relationship between V_c and β. The figure is drawn by the plots placed in the longitudinal direction at 10-mm intervals. The circle plots in Fig. 3.25 are the conditions using increased thickness explosive showing quite stable experimental conditions through the welding line. In addition, the triangle plots are the condition using fixed-thickness explosive showing decrease of both the velocity V_c and angle β leading to a lower energetic condition toward the end side of the plate. As shown in Fig. 3.24, it is found that the underwater shock wave is reflected after it acts on the surface of the cover plate, and the underwater shock pressure is about 1 GPa all through the length [57].

Based on the development of the assembly for underwater explosive welding, various combinations of materials have been welded successfully. For example, Hokamoto et al. [50] achieved the welding of a thin aluminum plate with a zirconia ceramic plate. Before that, only spot joining of a metal onto a ceramic block using a small amount of explosive, milligram order, had been reported [56], which was welded based on an anchoring effect of the metal deformed on the rough surface of a ceramic plate. In addition, the use of an underwater shock wave enables the welding of a wide area, such as plate joining [50]. The welding interface shows a planar structure due to the difference of the strength or hardness of materials inducing a metal jet only in the softer component, aluminum [50]. Once a wavy interface is formed, cracks are expected in the ceramic plate. For the welding of other brittle materials, Sun et al. [51] succeeded in welding a thin copper plate onto tool steel showing a clear wavy interface even by using tool steel after heat treatment for hardening. The decrease of kinetic energy may suppress the formation of cracks in such brittle metals.

Fig. 3.25 shows scanning electron micrographs of the interface for a thin tungsten plate welded onto a copper plate [63]. The welding of high melting point metal (W) with highly heat conductive metal (Cu) is believed quite attractive for practical applications. As seen in Fig. 3.25, both metals are well deformed, making waves and ripples mixing both components together. In addition, tungsten was well deformed, although the elongation at room temperature was not expected to be high and it is normally known as a brittle metal. The explosive welding of this combination using a conventional method is not possible due to the brittle fractures of tungsten [63]. The welding of W/Cu is difficult depending upon the sheets employed (case by case) and the thickness of the tungsten plate, and it is possible to have a sample without such brittle fractures in the tungsten by the use of a thin tungsten plate, which enables easier bending deformation. Fig. 3.26 shows the interface of W/Cu recovered under the inclination angle of explosive α set at 15 degrees [63]. When decreasing the setup angle, the formation of a wavy interface is not promoted and it increases the chance of trapping the metal jet spread ahead of the collision point. The flat interfacial layer found in the interface

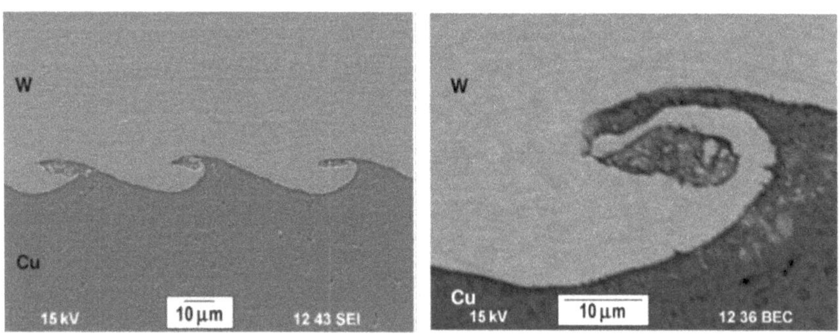

A Scanning electron microscope image of the welded interface

B a representative image of the vortex

Fig. 3.26 ***Interfacial waves formed in 0.1-mm-thick W welded onto a copper base plate.*** (A) Scanning electron microscopy image of the welded interface. (B) A representative image of the vortex. *(From P. Manikandan, J. O. Lee, K. Mizumachi, A. Mori, K. Raghukandan, and K. Hokamoto. Underwater explosive welding of thin tungsten foils and copper.* Journal of Nuclear Materials *418 (2011) 281–285).*

is composed of a fine mixture of tungsten and copper, because there is no intermetallic compound for the combination as found in its phase diagram, which is known as a miscibility gap alloy, not making a solid solution and not inducing an intermetallic compound. Normally, explosive welding is prohibited by the formation of brittle intermetallics, but the combination does not harm the quality of welding and it is sometimes considered better to play the role of a functionally grated layer. The same kinds of combinations, such as Nb/Cu and Ta/Cu, have been welded successfully also using an underwater explosive welding technique [26].

The welding of metals with low melting points is also difficult. Excessive melting under excessively energetic conditions leads to easy debonding. Aluminum and magnesium and their alloys are considered important for engineering applications, and some attempts have been made for Al/Mg by changing the parameters of underwater explosive welding [63]. For the welding of Al/Mg, the decrease in the melting point when mixing the two elements, Al and Mg, is induced, which may cause excessive melting, harming the welding. Therefore, the condition should be chosen carefully. The welding of titanium and magnesium is a more difficult combination due to the reactivity of both components. For this combination, the use of an intermediate magnesium alloy (AZ31) plate between the titanium flyer and the AZ31 base plates is effective to decrease the kinetic energy lost by collision, which can improve the quality of welding [15]. The welding interface sometimes induces a molten layer similar to the case for W/Cu (Fig. 3.27), and the present case is also a miscibility gap alloy combination, so the bonding is not harmed much, even with a fine mixed layer of Ti and Mg.

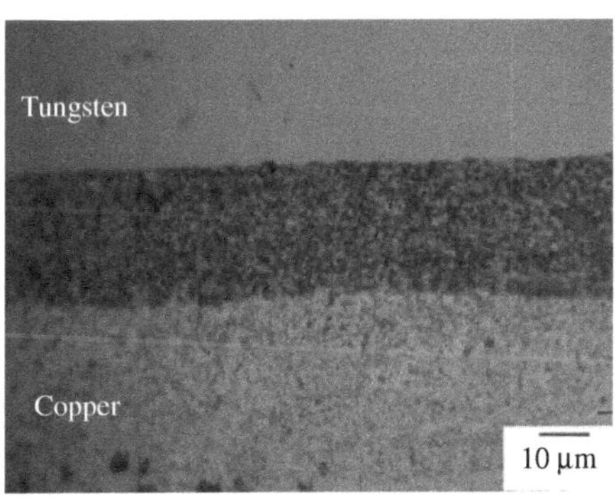

Fig. 3.27 ***Optical micrograph of the 0.1-mm tungsten and copper welded at a preset angle of 15 degrees.*** *(From P. Manikandan, J. O. Lee, K. Mizumachi, A. Mori, K. Raghukandan, and K. Hokamoto. Underwater explosive welding of thin tungsten foils and copper.* Journal of Nuclear Materials *418 (2011) 281–285).*

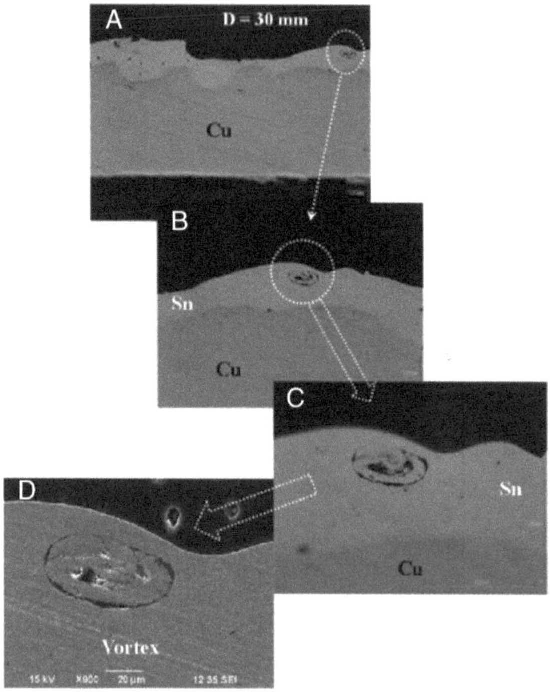

Fig. 3.28 ***Optical and scanning electron microscopy images of a vortex observed under the swirl of Sn.*** From A. M. Satyanarayan, M. Nishi, and K. Hokamoto. Underwater shock wave weldability window for Sn-Cu plates. *Journal of Materials Processing Technology* 267 (2019) 152–158.

The welding of a thin tin plate onto a copper base plate was also investigated by Satyanarayan et al. [61,64]. The welding was not extremely difficult with a wavy interface, but it is interesting to note that the intense dynamic flow at the interface moved deeply into the tin plate, suggested as a "vortex," as seen in Fig. 3.28 [64].

So far, the use of an inclined setup of explosive has been discussed, but it is possible to use a parallel setup when isolating an ANFO-based explosive from water. Some experiments have been carried out by Inao et al. [65], and it has been reported that it is also possible to weld various metals. Inao et al. [65] also suggested the possibility of the use of gelatin as a pressure transmitting medium instead of water. The responses to the shock wave in water and gelatin are considered quite similar based on optical observation experiments [66].

3.5 Conclusions

Explosive welding is widely industrialized all over the world and is the most important technology in the field of explosive materials processing. Current industrial demands are moving toward the welding of difficult-to-weld material combinations. Therefore, the information of the interfacial microstructure is important to improve the quality of welding, and there are some quite unique microscopic features found at the interface due to the drastic change of the state of materials, such as rapid cooling. The new explosive welding technique, using an underwater shock wave, makes it possible to modify the condition of explosive welding, and it is the hope of the authors to explore the industrial application of underwater explosive welding in the future.

References

[1] B. Crossland, Explosive Welding and Its Application, Oxford University Press, Oxford, UK, 1982.

[2] R.H. Wittman, The influence of collision parameters on the strength and microstructure of an explosion welded aluminum alloy, Proceedings of the 2nd International Symposium on Use of an Explosive Energy in Manufacturing Metallic Materials, 1973 153–168.

[3] A.A. Deribas, V.A. Simonov, I.D. Zakcharenko, Investigation of the explosive parameters for arbitrary combinations of metals and alloys, Proceedings of the 5th International Conference on High Energy Rate Fabrication, Denver, CO, June 24–26, 1975 24 4.1.1–4.1.

[4] A.A. Deribas, V.M. Kudinov, F.I. Matveenkov, V.A. Simonov, Explosive welding, Combustion, Explosion, and Shock Waves 3 (1) (1967) 111–118.

[5] G.R. Cowan, A.H. Holtzman, Flow configuration in colliding plates: explosive welding, Journal of Applied Physics 34 (1963) 928–939.

[6] G.R. Cowan, O.R. Bergman, A.H. Holtzman, Mechanism of bond zone wave formation in explosion-clad metals, Metallurgical and Materials Transactions 2 (1971) 3145–3155.

[7] H. Keno, Explosive welding of dissimilar material for ship [In Japanese.], Journal of the Japan Welding Society 87 (1) (2018) 76–79.

[8] V.I. Lysak, S.V. Kuzmin, Explosive Welding, Mashinostroenie, Moscow, Russia, 2005 [In Russian.].

[9] Y.Y. Émurlaeva, I.A. Bataev, Q. Zhou, D.V. Lazurenko, I.V. Ivanov, P.A. Riabinkina, et al., Welding window: comparison of Deribas' and Wittman's approaches and SPH simulation results, Metals 9 (12) (2019) 1323.

[10] I.A. Bataev, D.V. Lazurenko, S. Tanaka, K. Hokamoto, A.A. Bataev, Y. Guo, et al., High cooling rates and metastable phases at the interfaces of explosively welded materials, Acta Materialia 135 (2017) 277–289.

[11] H. Paul, M.M. Miszczyk, A. Galka, R. Chulist, Z. Szulc, Microstructural and chemical composition changes in the bonding zone of explosively welded sheets, Archives of Metallurgy and Materials 64 (2) (2019) 683–694.

[12] I.A. Bataev, T.S. Ogneva, A.A. Bataev, V.I. Mali, M.A. Esikov, D.V. Lazurenko, et al., Explosively welded multilayer Ni-Al composites, Materials and Design 88 (2015) 1082–1087.

[13] J. Nishiwaki, S. Kumai, Simulation of local melting zone formation in Cu/Al explosive welded joint [In Japanese.], Quarterly Journal of the Japan Welding Society 34 (4) (2016) 274–284.
[14] G.H.S.F.L. Carvalho, I. Galvão, R. Mendes, R.M. Leal, A. Loureiro, Influence of base material properties on copper and aluminium–copper explosive welds, Science and Technology of Welding and Joining 23 (2018) 501–507.
[15] M. Hammerschmidt, H. Kreye, Microstructure and bonding mechanism in explosive welding, in: M.A. Meyers (Ed.), Shock Waves and High-Strain-Rate Phenomena in Metals Concepts and Applications, Springer, New York, NY, 1981, pp. 961–973.
[16] M. Yang, H. Ma, Z. Shen, Z. Huang, Q. Tian, J. Tian, Dissimilar material welding of tantalum foil and Q235 steel plate using improved explosive welding technique, Materials and Design 186 (2020) 108348.
[17] J. Nishiwaki, S. Kumai, Simulation of cooling and intermediate layer formation processes at explosive welded Cu/Al interface, Quarterly Journal of the Japan Welding Society 35 (3) (2017) 111–121.
[18] Z.L. Zhang, M.B. Liu, Numerical studies on explosive welding with ANFO by using a density adaptive SPH method, Journal of Manufacturing Processes 41 (2019) 208–220.
[19] I.A. Bataev, Structure of explosively welded materials: experimental study and numerical simulation. [In Russian.], Metal Working and Material Science 4 (77) (2017) 55–67.
[20] I.A. Bataev, A.A. Bataev, V.I. Mali, V.G. Burov, E.A. Prikhod'ko, Formation and structure of vortex zones arising upon explosion welding of carbon steels, Physics of Metals and Metallography 113 (3) (2012) 233–240.
[21] A.A. Popov, L.E. Popova, Isothermal and Thermokinetic Diagrams of Decomposition of Supercooled Austenite: A Handbook of Heat Treatment Specialist, Metallurgia, Moscow, Russia, 1961 [In Russian.].
[22] E.B. Makarova, M.A. Esikov, A.S. Gontarenko, T.S. Sameishcheva, Structural changes in VT1-O titanium induced by deformation during explosion welding, Russian Metallurgy (Metally) (2012) 826–830 2012.
[23] A. Nassiri, T. Abke, G. Daehn, Investigation of melting phenomena in solid-state welding processes, Scripta Materialia 165 (2019) 61–66.
[24] I.N. Maliutina, V.I. Mali, I.A. Bataev, A.A. Bataev, M.A. Esikov, A.I. Smirnov, et al., Structure and microhardness of Cu-Ta joints produced by explosive welding, Scientific World Journal 2013 (5) (2013) 256758.
[25] B.A. Greenberg, M.A. Ivanov, A.M. Patselov, Y.P. Besshaposhnikov, The processes of fragmentation, intermixing and fusion upon explosion welding, AASRI Procedia 3 (2012) (2012) 66–72.
[26] P. Parchuri, S. Kotegawa, H. Yamamoto, K. Ito, A. Mori, K. Hokamoto, Benefits of intermediate-layer formation at the interface of Nb/Cu and Ta/Cu explosive clads, Materials and Design 166 (2019) 107610.
[27] B.A. Greenberg, M.A. Ivanov, A.V. Inozemtsev, A.M. Patselov, M.S. Pushkin, A.M. Vlasova, Microheterogeneous structure of local melted zones in the process of explosive welding, Metallurgical and Materials Transactions 46 (8) (2015) 3569–3580.
[28] P.K. Parchuri, S. Kotegawa, K. Ito, H. Yamamoto, A. Mori, S. Tanaka, et al., Cladding of a crack-free W plate on Cu plates using explosive welding at higher collision velocity with lower collision angle, Results in Materials 5 (2020) 100023.
[29] M. Nishida, A. Chiba, Y. Honda, J. Hirazumi, K. Horikiri, Electron microscopy studies of bonding interface in explosively welded Ti/steel clads, ISIJ International 35 (2) (1995) 217–219.
[30] M. Nishida, A. Chiba, Y. Morizono, M. Matsumoto, T. Murakami, A. Inoue, Formation of nonequilibrium phases at collision interface in an explosively welded Ti/Ni clad, Materials Transactions, JIM 36 (11) (1995) 1338–1343.
[31] I.A. Bataev, D.V. Lazurenko, Y.N. Malyutina, A.A. Nikulina, A.A. Bataev, O.E. Mats, et al., Ultrahigh cooling rates at the interface of explosively welded materials and their

effect on the formation of the structure of mixing zones, Combustion, Explosion, and Shock Waves 54 (2) (2018) 238–245.
[32] D.V. Lazurenko, I. A. Bataev, V. I. Mali, E. A. Lozhkina, M. A. Esikov, and V. A. Bataev. Structural transformations occurring upon explosive welding of alloy steel and high-strength titanium. Physics of Metals and Metallography 119 (5) 469–476.
[33] K.F. Kobayashi, T. Awazu, P.H. Shingu, New metastable phases in the Al-Mg system, Journal of Materials Science Letters 6 (1987) 781–783.
[34] U.D. Kulkarni, G.K. Dey, R.T. Savalia, S. Banerjee, A near icosahedral quasiperiodic phase in rapidly solidified $Mg_{32}Al_{49}$, Philosophical Magazine Letters 64 (3) (1991) 169–174.
[35] P. Donnadieu, H.L. Su, A. Proult, M. Harmelin, G. Effenberg, F. Aldinger, From modulated phases to a quasiperiodic structure with a cubic point group and inflation symmetry, Journal de Physique I 6 (1996) 1153–1164.
[36] L. Elcoro, J.M. Perez-Mato, Cubic superspace symmetry and inflation rules in metastable MgAl alloy, European Physical Journal B 7 (1999) 85–89.
[37] D.V. Lazurenko, I.A. Bataev, V.I. Mali, A.A. Bataev, I.N. Maliutina, V.S. Lozhkin, et al., Explosively welded multilayer Ti-Al composites, structure and transformation during heat treatment, Materials and Design 102 (2016) 122–130.
[38] H. Paul, J. Morgiel, T. Baudin, F. Brisset, M. Prazmowski, M. Miszczyk, Characterization of explosive weld joints by TEM and SEM/EBSD, Archives of Metallurgy and Materials 59: (2014) 1129–1136.
[39] H. Paul, M.M. Miszczyk, R. Chulist, M. Prażmowski, J. Morgiel, A. Gałka, et al., Microstructure and phase constitution in the bonding zone of explosively welded tantalum and stainless steel sheets, Materials and Design 153 (2018) 177–189.
[40] H. Paul, W. Skuza, R. Chulist, M. Miszczyk, A. Gałka, M. Prażmowski, et al., The effect of interface morphology on the electro-mechanical properties of Ti/Cu clad composites produced by explosive welding, Metallurgical and Materials Transactions A 51 (2020) 750–766.
[41] G. Zu, X. Li, J. Zhang, H. Zhang, Interfacial characterization and mechanical property of Ti/Cu clad sheet produced by explosive welding and annealing, Journal of Wuhan University of Technology—Materials Science Edition 30 (6) (2015) 1198–1203.
[42] A. Inoue, Stabilization of metallic supercooled liquid and bulk amorphous alloys, Acta Materialia 48 (1) (2000) 279–306.
[43] I.A. Bataev, K. Hokamoto, H. Keno, A.A. Bataev, I.A. Balagansky, A.V. Vinogradov, Metallic glass formation at the interface of explosively welded Nb and stainless steel, Metals and Materials International 21 (4) (2015) 713–718.
[44] A.R. Yavari, K. Georgarakis, J. Antonowicz, M. Stoica, N. Nishiyama, G. Vaughan, et al., Crystallization during bending of a Pd-based metallic glass detected by x-ray microscopy, Physical Review Letters 109(8) (2012) 085501.
[45] Z.L. Mao, H. Chen, W.K. Wang, Formation of bulk metallic glass $Zr_{60}Ni_{20}Al_{20}$ by high-pressure quenching, Journal of Materials Science Letters 12 (21) (1993) 1729–1730.
[46] H. Kato, A. Inoue, J. Saida, Influences of hydrostatic pressure during casting and Pd content on as-cast phase in Zr-Al-Ni-Cu-Pd bulk alloys, Applied Physics Letters 85 (12) (2004) 2205–2207.
[47] X.R. Liu, S.M. Hong, S.J. Lü, R. Shen, Preparation of $La_{68}Al_{10}Cu_{20}Co_2$ bulk metallic glass by rapid compression, Applied Physics Letters 91 (2007) 081910.
[48] I.A. Bataev, S. Tanaka, Q. Zhou, D. Lazurenko, A.J. Junior, A.A. Bataev, et al., Towards better understanding of explosive welding by combination of numerical simulation and experimental study, Materials and Design 169 (2019) 107649.
[49] M.A. Meyers, Dynamic Behavior of Materials, John Wiley & Sons, New York, NY, 1994.
[50] K. Hokamoto, M. Fujita, H. Shimokawa, H. Okugawa, A new method for explosive welding of Al/ZrO_2 joint using regulated underwater shock wave, Journal of Materials Processing Technology 85 (1999) 175–179.

[51] W. Sun, X. Li, H. Yan, K. Hokamoto, Effect of initial hardness on interfacial features in underwater explosive welding of tool steel SKS3, Journal of Materials Engineering and Performance 23 (2) (2014) 421–428.
[52] K. Hokamoto, U. Ujimoto, M. Fujita, Basic characteristics of the explosive welding technique using underwater shock wave and its possibilities, Materials Transactions 45 (9) (2004) 2897–2901.
[53] S.P. Marsh, LASL Shock Hugoniot Data, University of California Press, Berkeley, 1980.
[54] R.H. Cole, Underwater Explosions, Princeton University Press, Princeton, NJ, 1948.
[55] D.J. Vigueras, O.T. Inal, A. Szecket, Explosive welding of an amorphous ribbon to a mild steel substrate. In: Metallurgical Applications of Shock-Wave and High-Strain-Rate Phenomena, Decker, New York, NY, 1986, pp. 927–942.
[56] J.W. Shaffer, B.H. Cranston, G. Krauss, Explosive bonding of metal foils to high alumina ceramic substrates, *Proceedings of the 5th International Conference on High Energy Rate Fabrication*, Denver, CO, June 24--26, 1975, pp. 28 4.12.1–4.12.
[57] M. Fujita, K. Hokamoto, A. Mori, Welding conditions for underwater explosive welding of amorphous film onto mild steel [In Japanese.], Proceedings of the Materials and Processing Conference, 15, 2007 173–174.
[58] K. Hokamoto, K. Nakata, A. Mori, S. Tsuda, T. Tsumura, A. Inoue, Dissimilar material welding of rapidly solidified foil and stainless steel plate using underwater explosive welding technique, Journal of Alloys and Compounds 472(1-2) (2009) 507–511.
[59] P. Manikandan, J. Lee, K. Mizumachi, S.H. Ghaderi, A. Mori, K. Hokamoto, Transition joints of aluminum and magnesium alloy made by underwater explosive welding technique, Materials Science Forum 706 (709) (2012) 757–762.
[60] M. Ahasan Habib, H. Keno, R. Uchida, A. Mori, K. Hokamoto, Cladding of titanium and magnesium alloy plates using energy-controlled underwater three layer explosive welding, Journal of Materials Processing Technology 217 (2015) 310–316.
[61] S.T. Satyanarayan, A. Mori, K. Hokamoto, Welding of Sn and Cu plates using controlled underwater shock wave, Journal of Materials Processing Technology 245: (2017) 300–308.
[62] Century Dynamics Inc. *AUTODYN Theory Manual Revision* 4.3, Century Dynamics Inc, Concord, CA, 2003.
[63] P. Manikandan, J.O. Lee, K. Mizumachi, A. Mori, K. Raghukandan, K. Hokamoto, Underwater explosive welding of thin tungsten foils and copper, Journal of Nuclear Materials 418 (2011) 281–285.
[64] A.M. Satyanarayan, M. Nishi, K. Hokamoto, Underwater shock wave weldability window for Sn-Cu plates, Journal of Materials Processing Technology 267 (2019) 152–158.
[65] D. Inao, A. Mori, S. Tanaka, K. Hokamoto, Explosive welding of thin aluminum plate onto magnesium alloy plate using a gelatin layer as a pressure-transmitting medium, Metals 10 (1) (2020) 106.
[66] D. Inao, S. Tanaka, T. Yamashita, K. Hokamoto, Visualization of shockwave behavior in water and gelatin, Measurement 148 (2019) 106929.

CHAPTER 4

Behavior of materials under extremely high-velocity oblique impact

Ivan Bataev[a], Akihisa Mori[b], Kazuyuki Hokamoto[c]
[a]Novosibirsk State Technical University, Russia
[b]Sojo University, Kumamoto, Japan
[c]Professor, Institute of Industrial Nanomaterials, Kumamoto University, Kumamoto, Japan

4.1 Introduction

In recent years, the behavior of solids during high-velocity oblique collisions has attracted considerable interest from the scientific community. There are several reasons for this. First, the oblique collision of plates is of great importance for various high-velocity impact welding technologies. Such technologies, for example, include explosive welding, magnetic pulse welding, and vaporizing foil actuator welding, which are used in industry for the joining of dissimilar materials. Second, this topic is of interest when creating protective materials. In particular, oblique collision at hypersonic velocities is relevant for creating screens that protect spacecraft from microscopic dust. Third, oblique collision is the basic phenomenon of shaped charges, which have practical applications in both military and civilian spheres. In the latter case, it is used for the demolition of large-sized constructions and hole punching for the oil and gas industry.

All of these beforementioned technologies have several common phenomena. Depending on the materials and the regime of impact (collision angle and impact velocity), solid bodies may rebound or weld to each other. In many cases, interface instabilities, localized melting, and intermixing of materials are observed between the colliding plates. In addition, oblique collision at extremely high velocities in many cases is accompanied by the formation of a cumulative jet (or plasma cloud) in front of the collision point.

Many different approaches can be used to study the behavior of materials during high-velocity impact. However, the most informative is the visualization of the collision process using a laboratory powder gun equipped with a high-speed video camera and numerical simulation of the process. The advantage of powder guns is the ability to accurately measure

Explosion, Shock-Wave and High-Strain-Rate Phenomena of Advanced Materials
DOI: 10.1016/C2019-0-03954-4

the impact velocity and impact angle. In addition, the detonation products do not prevent observation of the collision region. As a rule, powder gun experiments are carried out in a vacuum, which makes it possible to simplify the analysis of the results.

Throughout almost the entire history of research in the field of high-velocity collisions, various methods of numerical simulation have been used to analyze them. Finite element analysis using Lagrangian and Eulerian methods has proven their high efficiency and regularly are used in study of shock wave processes. Over the past decade, increasing attention has been paid in the literature to simulation using the smooth particle hydrodynamics (SPH) method. This is largely because this method is well suited to the analysis of extremely large plastic deformations and well reproduces the experimentally observed interfaces of plates welded by impact. In particular, it reproduces well the instabilities arising at the boundaries in the form of waves and molten microvolumes with a specific shape and location. For this reason, in this section, mainly the results of simulation using the SPH method will be presented.

In addition, important conclusions about the behavior of materials under conditions of high-velocity oblique impact can be made by metallographic analysis of the samples after completion of the process. For example, the analysis of microstructures formed as a result of high strain rate interaction allows one to make indirect conclusions about the temperatures and pressures that occur in the collision process. In the ideal case, for a more complete understanding of the phenomena occurring during a high-velocity collision, it is advisable to use a combination of all preceding approaches.

This chapter mainly presents the results of our own experimental studies and numerical simulations of oblique collisions using a powder gun. Most of the experiments were conducted at the Institute of Industrial Nanomaterials (formerly the Institute of Pulsed Power Science), Kumamoto University. From our point of view, the results obtained are of interest both for a fundamental understanding of high-velocity collision phenomena, and can be of practical value particularly for choosing the optimal collision regimes when implementing explosion welding technology.

4.2 Observation of high-velocity oblique collisions using a high-speed video camera

The experimental setup for the observation of high-velocity collision of plates is shown in Fig. 4.1. For this kind of experiment, we use a

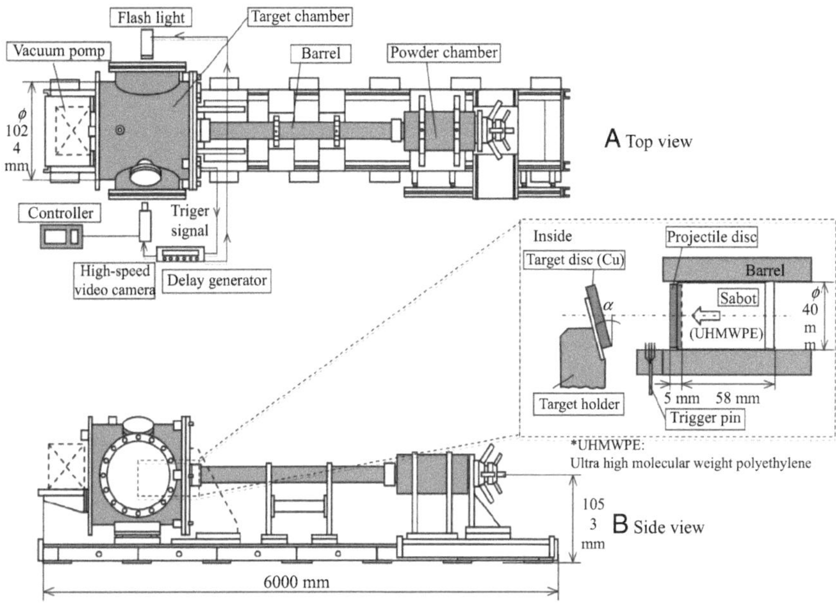

Fig. 4.1 ***Scheme of a laboratory single-stage powder gun installed at the Institute of Industrial Nanomaterials, Kumamoto University. The enlarged image shows the layout of the projectile and target samples in the experimental chamber.*** **UHMWPE, *Ultra-high-molecular-weight polyethylene*.**

From A. Mori, S. Tanaka, and K. Hokamoto. Optical observation of metal jet generated by high speed inclined collision. In Proceedings of SPIE—the 31st International Congress on High-Speed Imaging and Photonics, Osaka, Japan 2016, November 7–10 (2016), 103281Q-1–103281Q-4.

laboratory single-stage powder gun installed at the Institute of Industrial Nanomaterials, Kumamoto University. The gun has a caliber of 40 mm and allows one to shoot sabots weighing from 50 to 200 g. As a rule, the sabots are made of ultra-high-molecular-weight polyethylene (UHMWPE). The experimental samples are usually glued to the projectiles. UHMWPE has a low coefficient of friction with steel, and it is also quite soft and easy to machine, and for this reason it is well suited for our experimental purposes. Black powder or smokeless powder are used to accelerate the sabot. With this experimental setup, it is possible to use powder charges weighing from 4 to 100 g, which ensures a projectile speed of 200 to 1500 m/s. Powder ignition is initiated by a special electric fuse. The gun was mounted on rails so that it can be pushed toward the experimental chamber, which was equipped with two large windows made of polymethylmethacrylate (PMMA), through which it was convenient to observe the collision process using a high-speed video camera. Before starting the experiments, the chamber and the barrel were evacuated using a forevacuum pump to a pressure of 3×10^{-3} Pa.

The collision process was observed using high-speed video cameras, an ultra-high-speed HyperVision HPV-1 camera (Shimazu Co. Ltd.) that allows one to shoot up to 100 frames with a frequency ranging from 30 to 1,000,000 frames per second and a minimum exposure time of 0.125 μs. The camera is equipped with an 8-bit CCD matrix with a resolution of 312 × 260 pixels. It can be connected to any lens using the Nikon F bayonet mount. The camera requires a powerful light source, which can be a metal halide lamp or an arc flash lamp.

Since the maximum number of frames is limited, and the phenomena occurring during the collision are extremely transient, it was important to use a trigger signal that allows one to start recording at the most important moment of the experiment. For instance, Fig. 4.1 shows a trigger wire, the contacts of which are closed during interaction with the sample and indicate the beginning of the recording moment. This approach allows one to precisely capture the moment of impact despite the limitations in the number of frames.

Several examples of this setup application are described in the following. In the first case, the welding of two copper plates was studied, and the second case was the welding of copper and aluminum plates [1]. These experiments were carried out to reproduce the conditions typical of explosion welding and to visualize the phenomena occurring near the collision point.

In both cases, copper disks with a diameter of 32 mm and a thickness of 5 mm were used as targets. The target was mounted on a special holder at an angle of 15 or 20 degrees relative to the flyer sample. Copper or aluminum plates with a diameter of 32 mm and a thickness of 5 mm were used as flyer plates. They were glued to UHMWPE sabots. The impact velocity was in the range of 400 to 650 m/s. Collision angles and impact velocities were selected using precalculated welding windows to ensure reliable bonding of materials in the regimes, which are typical of explosion welding.

The images recorded after 4, 8, 12, and 16 μs from the beginning of the impact are shown in Fig. 4.2 [1]. Both during the collision of similar materials and during the collision of dissimilar materials, the formation of a metal jet was observed. In the case of copper to copper collision (Fig. 4.2A), the jet was in the form of a cloud or a spray under all impact conditions. In the case of aluminum and copper plates collision, a bright glow of the jet was observed (Fig. 4.2B). The same luminescence was observed during collision of plates made of copper and magnesium alloy AZ31 [1]. Since

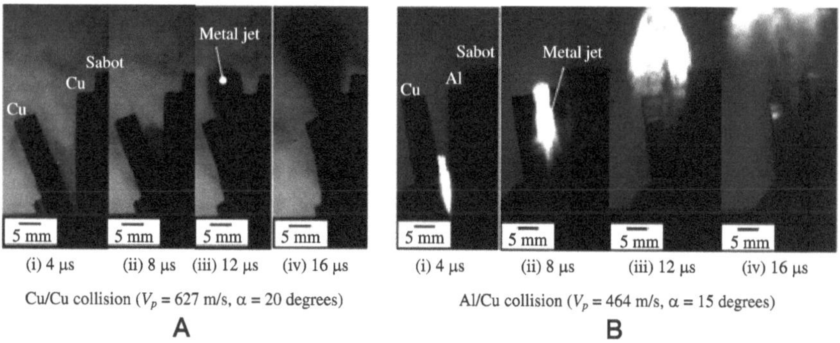

Fig. 4.2 ***Images of the jet formation during collision of copper to copper (A) and of aluminum to copper (B).***
From A. Mori, S. Tanaka, and K. Hokamoto. 2016. Optical observation of metal jet generated by high speed inclined collision. In Proceedings of SPIE—the 31st International Congress on High-Speed Imaging and Photonics, Osaka, Japan, November 7–10, 2016, 103281Q-1–103281Q-4.

aluminum and magnesium are extremely reactive materials, their luminescence could have occurred due to an exothermic chemical reaction. For the cases presented in Fig. 4.2, the velocity of the jet was measured. It was equal to 3537 and 4092 m/s for the two cases shown in Fig. 4.2A and B, respectively. The velocity of the jet was approximately two times the velocity of the collision point.

4.3 Numerical simulation of high-velocity oblique collision

Numerical simulation of high strain rate phenomena is one of the most powerful and affordable tools for scientists studying the dynamic behavior of materials. The active development of numerical methods for solving hydrodynamic problems began in the 1960s. Computer programs developed to solve the equations of continuum mechanics were then called *hydrocodes*, although they were already complex enough to move away from the purely hydrodynamic concept of materials behavior and use various models of strength and fracture of solids [2]. The application of these methods to oblique collisions was one of the priority areas of research, primarily for the analysis of shaped charges. Numerical simulation methods have also been used for a long time to analyze high-velocity impact welding. For example, the study of Godunov et al. [3], published in the 1970s, was devoted to numerical simulation of high-velocity impact welding of copper plate to steel plate. The impact velocity was 500 m/s and the impact angle was 10 degrees. They showed that the highest pressure at the collision point was about 15 GPa. At the interface behind the collision point, they observed formation of shallow waves. Nevertheless, these waves were significantly different from those usually found in explosively welded materials. The apparent formation of a jet in front of the collision point was not observed at that time. It should be kept in mind that the resolution of the model was rather "rough," as the computational domain consisted in total only of 1500 nodes [4].

Mouasavi et al. [5] claimed they were the first to reproduce jet formation in a simulation of explosion welding of titanium and stainless steel. They used ANSYS AUTODYN 2D software. For explosives, the Williamsburg equation of state was used. For air, the ideal gas equation of state was used. They observed that some material was squeezed in front of the collision point, however it looked different from the jet observed in experiments. They did not report the formation of waves at the interface, but from the images shown in the publication, it can be concluded that the simulation predicted flat interfaces.

It should be noted that the work of Grignon et al. [6] was submitted to the International Journal of Impact Engineering several months earlier than the work of Mousavi et al. [5] but appeared in print later. It simulated explosion welding of aluminum alloy pipes using an approach similar to that of Mousavi et al. [5]. Grignon et al. also observed the formation of some kind of jet; however, the formation of waves was not observed despite the fact that they were found in experimental studies. The formation of waves using the Eulerian solver was reported by Mousavi and Al-Hassani [7] several years later in a study on simulation of collisions at high velocities. A fairly good correlation was achieved between experiment and simulation predictions.

In the analysis of explosive welding, the simulation of all of the phenomena accompanying this process (detonation of explosives, expansion of detonation products, acceleration of a flyer plate, deformation at the interface, etc.) is a costly issue in terms of computational resources. Correct simulation of the detonation process and plate acceleration requires that the geometric dimensions of the computational domain should be large enough so that the stationary parameters of the flyer plate are established, whereas sufficient calculation accuracy is ensured with a relatively rough spatial resolution. At the same time, the analysis of subtle phenomena occurring near the collision point (the formation of waves, vortex zones, jets, etc.) requires a high-resolution simulation. Thus, the total calculation time for reproducing all of the details when using modern personal computers can be measured in weeks or even months. For this reason, it is reasonable to split the simulation of the collision process into two parts. The first part is devoted to the analysis of detonation and plate acceleration by detonation products, and a relatively coarse resolution of the computational grid is used to solve it. The obtained collision angles and impact velocities are used as initial conditions for the second problem, aimed at analyzing the process of plastic deformation during the collision. The second problem is solved for a computational domain that is small in geometric dimensions; however, the small size of the elements is used, which makes it possible to reproduce small geometric inhomogeneities, such as waves or jets, with high accuracy. Such an approach, for example, was used in the work of Bataev et al. [8], in which the process of underwater welding by explosion of niobium and stainless steel was studied. This approach allowed one to reproduce the explosion welding process within an acceptable time and with high accuracy. In particular, the shape of the interface and the formation of the jet were reproduced.

Significant progress in the simulation of materials' behavior in high-velocity collisions was achieved with the application of the SPH method. This method is well suited for simulating fast phenomena associated with extremely large strains, and for this reason, it is ideally suited for the analysis of deformation during high-velocity collisions. Probably the first study in which the SPH method was used to analyze high-velocity plate collision was published by Tanaka [9], in which he clearly observed the formation of waves and jets. Both of these phenomena were reproduced extremely plausibly. Li et al. [10] also observed wavy and flat interfaces, as well as jet formation during simulation of explosive welding using the SPH method. They demonstrated a good correspondence of the wave formation process during simulation with the mechanism of wave formation proposed by Bahrani et al. [11]. Currently, there is a rapid increase in the number of publications devoted to SPH simulation of welding involving high-velocity collisions. The method has been successfully used for simulation of explosive welding, welding in powder guns, and magnetic pulse welding [12]. A good correspondence was shown between the morphology of the emerging boundary observed experimentally and predicted by simulation for the following material combinations: W/Cu [13], Ti-6Al-4V/Al-1060 [14], steel/steel [15], Al/Zr-based bulk metallic glass [16], Cu/Ti [17,18], and steel/Al, among others.

As an example, illustrating the capabilities of the SPH method, we present here an analysis of the deformation of 6061-T6 aluminum alloy plates with a thickness of 6 mm. A significant number of collision regimes were simulated in the study of Émurlaeva et al. [19], and the obtained results were compared with experimental data for the same alloy obtained in the classical study of Wittman [20]. It can be noted that the simulation reproduces well the morphology of the interface (Fig. 4.3 [19]): at low collision point velocities, the flat interfaces form (Fig. 4.3A); with an increase in the collision velocity, the waves begin to form (Fig. 4.3B); and at high collision velocities, the formation of vortices (Fig. 4.3C) and even voids in them (Fig. 4.3D) is observed. Distortions of the reference grid indicate that the plastic strain (in particular, shear stain) increases significantly as one approaches the interface. It should be noted that the simulation results are in good agreement with the plastic strains experimentally measured near the interface in the work of Lysak and Kuzmin [21] (Fig. 4.4 [19]). The jet formation process is also extremely well reproduced, and the weldability window constructed from the calculated data is in good agreement with the weldability window for this alloy presented in the study of Wittman [20] (Fig. 4.5 [19]).

A visualization of the flow instability at the interface can be obtained by analyzing the streamlines constructed using the particle velocity field.

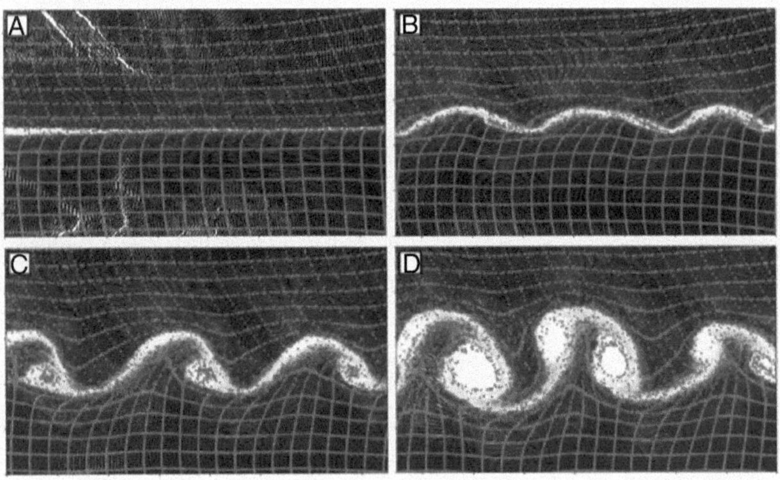

Fig. 4.3 ***The shape of the interfaces formed during high-velocity collision of 6061-T6 alloy plates predicted by the smooth particle hydrodynamics simulation. Impact angle: 25 degrees; contact point velocities: 1300 m/s (A), 2100 m/s (B), 2900 m/s (C), and 3700 m/s (D).***
From Y. Y. Émurlaeva, I. A. Bataev, Q. Zhou, D. V. Lazurenko, I. V. Ivanov, P. A. Riabinkina, S. Tanaka, and P. Chen. Welding window: comparison of Deribas' and Wittman's approaches and SPH simulation results. Metals 9(12)(2019) 1323.

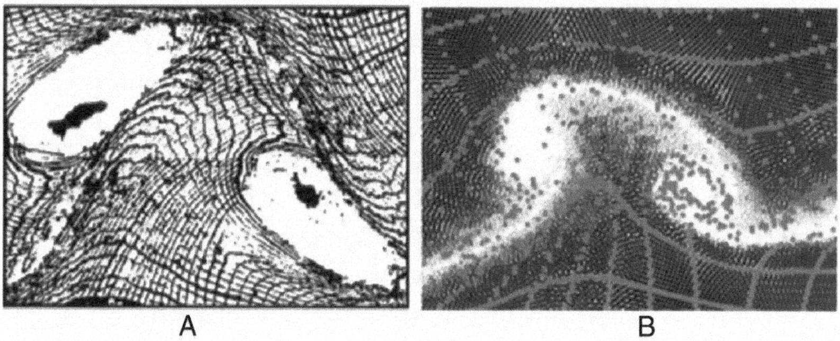

Fig. 4.4 ***Comparison of the experimental observation (A) and SPH simulation (B) of the plastic strains near the interface formed by high-velocity collision of plates of aluminum alloy 6061-T6.***
From Y. Y. Émurlaeva, I. A. Bataev, Q. Zhou, D. V. Lazurenko, I. V. Ivanov, P. A. Riabinkina, S. Tanaka, and P. Chen. Welding window: comparison of Deribas' and Wittman's approaches and SPH simulation results. Metals 9 (12)(2019) 1323.

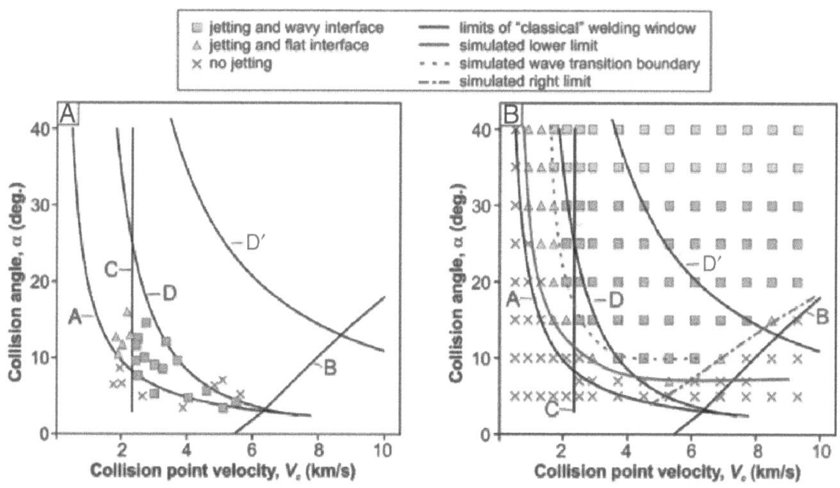

Fig. 4.5 ***(A) The welding window obtained using experimental data and semiempirical models. (B) A comparison of the results of Whitman with the results of the numerical simulation by Émurlaeva et al. Curves D and D' were calculated using Wittman's [20] and Deribas' [4] approaches, respectively.***
Part (A) from R. H. Wittman. The influence of collision parameters on the strength and microstructure of the explosion welded aluminum alloy. In Proceedings of the 2nd International Symposium on Use of an Explosion Energy in Manufacturing Metallic Materials, Marianske Lazne, Czech Republic, October 9–12(1973) 153–168. Part (B) from Y. Y. Émurlaeva, I. A. Bataev, Q. Zhou, D. V. Lazurenko, I. V. Ivanov, P. A. Riabinkina, S. Tanaka, and P. Chen. Welding window: comparison of Deribas' and Wittman's approaches and SPH simulation results. Metals 9 (12)(2019) 1323.

Fig. 4.6 shows three cases of impact of steel plates. If the collision point velocity is to the left of the lower welding limit, then jet formation does not occur (Fig. 4.6A). At a higher collision point velocity, an unstable flow is observed leading to the formation of waves, and the formation of a re-entrant jet also occurs (Fig. 4.6B). At a supersonic collision point velocity, the jet formation is again suppressed (Fig. 4.6A), which is in good agreement with the well-known ideas about the welding process. Using the vector velocity field, it is possible to show specific twisting that occurs in the vortex region during high-velocity impact welding (Fig. 4.7).

To simulate the mechanism of bonding during high-velocity impact welding at the atomic scale, the molecular dynamics method has also been widely used. For example, Kiselev and Mali [22] used this method to study the process of jet formation during high-velocity collision of nano-sized copper plates (with thicknesses from 6.2 to 24 nm). Despite the extremely small size of the simulated samples, the authors showed that the method reproduces all of the phenomena typical of impact of large-sized plates: at low-impact velocities, the material behaves elastically, and jet formation does not occur. With an increase in the Reynolds number to a certain critical value, jet formation is observed. At high Reynolds numbers, the simulated jet velocity is in agreement with that predicted by the ideal fluid model. In a collision at supersonic velocity, the formation of a jet occurs only at high collision angles. In addition, using the π-theorem, the authors compared the results obtained on such a small scale with the results obtained in experiments at the macro level and found good agreement between the

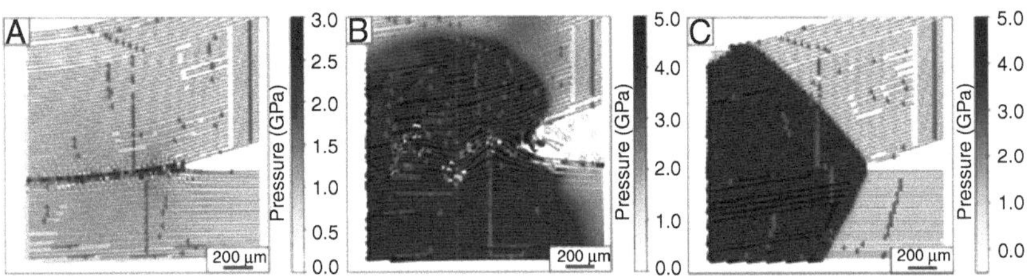

Fig. 4.6 ***Streamlines near the collision point during the oblique impact of steel plates. The coordinate system moves from left to right at the same velocity as the collision point: 1150 m/s (A), 3830 m/s (B), and 9580 m/s (C). The impact angle in all cases is 15 degrees. The background color indicates the pressure value (the color bar is shown to the right of each figure).***

From I. A. Bataev, S. Tanaka, Q. Zhou, D. V. Lazurenko, A. M. Jorge Jr., A. A. Bataev, K. Hokamoto, A. Mori, and P. Chen. Towards better understanding of explosive welding by combination of numerical simulation and experimental study. Materials and Design 169: 107649.

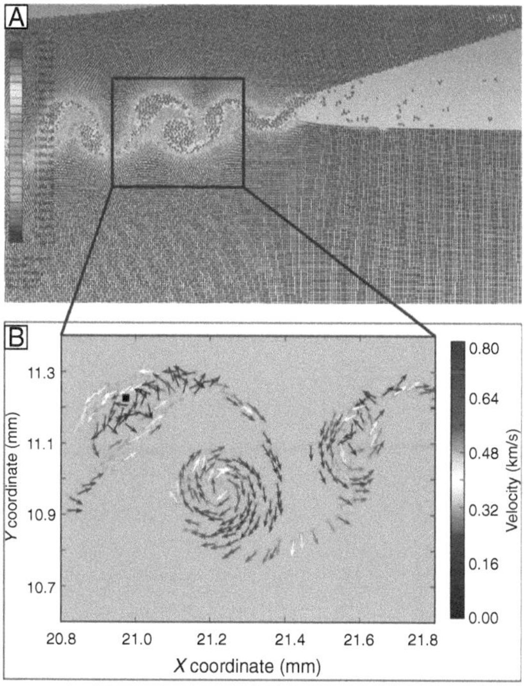

Fig. 4.7 ***Twisting of particles in vortex zones during the collision of steel plates.*** *From I. A. Bataev, S. Tanaka, Q. Zhou, D. V. Lazurenko, A. M. Jorge Jr., A. A. Bataev, K. Hokamoto, A. Mori, and P. Chen. Towards better understanding of explosive welding by combination of numerical simulation and experimental study. Materials and Design 169: 107649.*

experimental and calculated approaches. Later, Kiselev [23] used molecular dynamics to analyze the reasons leading to the formation of waves at the interface during high-velocity impact of plates. In accordance with his observations, wave formation is caused by self-sustaining oscillations in the high-pressure region. For the wavelength (λ) in aluminum, he obtained the following expression: $\lambda \approx 24\delta\sin^2(\gamma/2)$, where δ is the thickness of the flyer plate and γ is the angle between the plates. This expression correlates well with the experimental data of Deribas et al. [24].

In several studies, the molecular dynamics method was used to analyze diffusion processes occurring at the interface [25,26]. Feng et al. [26] used the molecular dynamics method to analyze the behavior of materials at three stages of explosive welding: loading, unloading, and cooling. Based on the simulation, it was concluded that the melting at the interface is not necessary for bond formation; however, under certain loading conditions, it occurs. In the absence of melting, mutual diffusion is substantially hindered.

In addition, the increase in pressure prevents the melting and the local melting of material is possible at the unloading stage. Rapid solidification leads to the formation of a nanograin structure at the interface, which has high tensile strength.

4.4 Microstructure of iron plates joined during oblique collision in powder gun

In this section, we present the phenomena occurring during high-velocity impact of steel plates by jointly analyzing the results of ultra-high-speed video camera observation, numerical simulation, and microstructural characterization. The collision experiments were carried out in the powder gun shown in Fig. 4.1.

Low-carbon steel disks (of 1-mm thickness and 38-mm diameter) made of 1006 alloy were used for the experiments. This type of steel was chosen because it is a convenient material, the structure of which is described under different types of deformation and thermal impacts. In addition, depending on the pressures and temperatures, several phase transformations may occur in steel (e.g., $\alpha \rightarrow \gamma$ and $\alpha \rightarrow \varepsilon$), by which it is possible to indirectly evaluate the conditions achieved in the collision process.

One of the disks was installed in the experimental chamber to provide a collision angle of 15 degrees. The second disk was glued to an UHMWPE cylinder, which played the role of a sabot. The sabot was accelerated in the barrel by a small charge of gunpowder. Before starting the experiments, the entire volume of the gun was evacuated to a pressure of 70 mbar or better. The collision process was recorded using an HPV-1 ultra-high-speed video camera. A 2-kW metal halide lamp (Iwasaki Electric Company, Ltd.) was used as a light source. Three experiments were carried out at the same impact angle (15 degrees) but different flyer plate velocities (542, 735, and 1000 m/s). The flyer plate velocity was measured using the video camera data.

The collision process recorded using the high-speed camera is shown in Fig. 4.8 [15]. Under all experimental conditions, a bright glow was formed in front of the contact zone of the plates from the first moments of collision (Fig. 4.8B). In various studies devoted to high-velocity impact welding, the material ejected in front of the contact point is referred to using slightly different terms. Frequently, such a material is called a *jet* or a *plasma jet*. However, the more cautious term that is sometimes used in the classical studies of Deribas [4] is a *cloud of particles*. In particular, Mali and

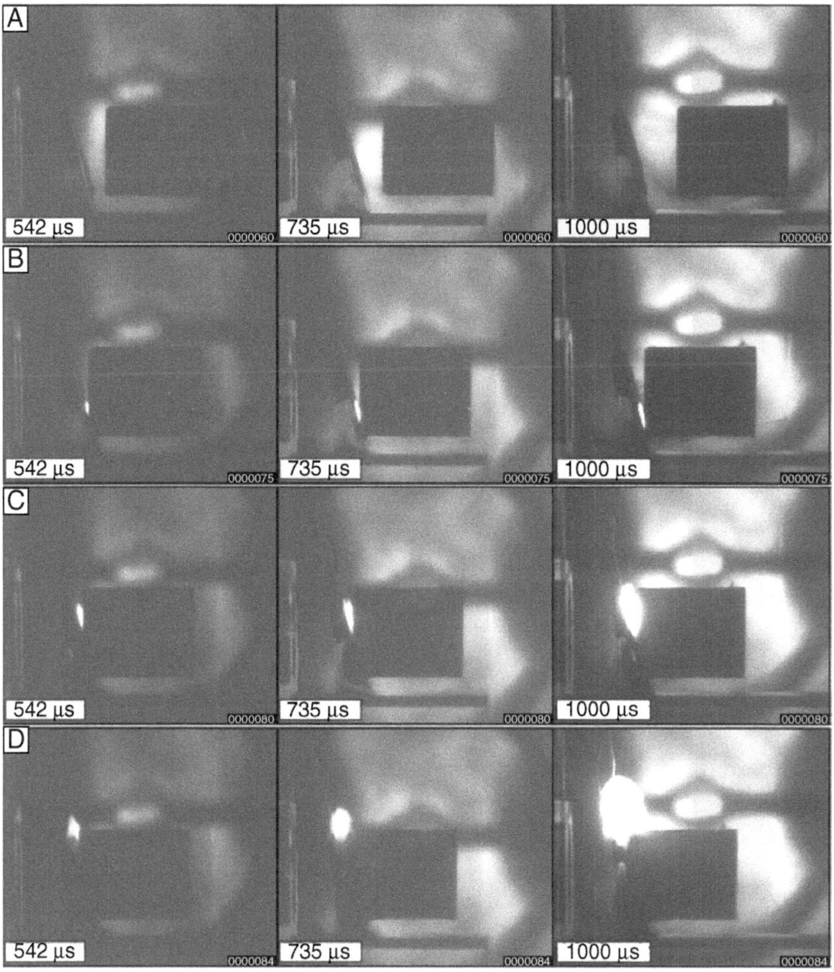

Fig. 4.8 ***High-speed camera images of the low-carbon steel plates' impact: a few microseconds before the collision (A), the beginning of the collision (B), 5 μs from the beginning of the collision (C), and 9 μs from the beginning of the collision (D).***
From I. A. Bataev, S. Tanaka, Q. Zhou, D. V. Lazurenko, A. M. Jorge Jr., A. A. Bataev, K. Hokamoto, A. Mori, and P. Chen. Towards better understanding of explosive welding by combination of numerical simulation and experimental study. Materials and Design 169: 107649.

Simonov [27] used high-speed X-ray photography to observe this phenomenon. They claimed that it is more rational to call the material moving in front of the collision point a *cloud*. This is because the shape of the material ejected in front of the collision point during explosive welding is "blurred" and differs significantly from the shape of the jets formed during the collapse of shaped charges. In addition, the material of the "cloud" has a significantly lower density in comparison with the welded plates. In this text, the term *cloud* will also be used to describe the glowing material moving in front of the collision point. The nature of the material in this cloud, however, deserves a separate discussion. First, it may be assumed that the shock-compressed gas propagates in front of the collision point. This shock-compressed gas can be heated to high temperatures, leading to its transition to the plasma state. In this work, the experiments were carried out in low vacuum and it is impossible to completely exclude the possibility of air plasma formation. However, from our point of view, a part of the observed glowing cloud is iron heated to a plasma state.

The welding of plates was observed under all collision conditions. The welded samples were investigated using microscopic techniques. In addition, the collision process was simulated using the commercial software ANSYS AUTODYN, and the simulation results were compared with the results of metallographic studies.

The cross sections of the samples along the collision point velocity vector are shown in Fig. 4.9 [15]. It follows that a wavy interface was formed in all cases. The wave size increased with the collision point velocity (Fig. 4.9A–C). This difference is especially noticeable between samples welded at impact velocities of 542 and 735 m/s. Fig. 4.9D–F presents enlarged images of the interface. The arrows in Fig. 4.10D show vortex zones, the material of which, as already noted in the previous chapter, was probably in a molten state. Vortex zones are located on the lateral sides of the waves; however, it is highly possible that thin melt zones appeared along the entire interface as a very thin and barely distinguishable layer.

Plastic strain of the material increases closer to the interface. The ferritic grains located at a considerable distance from the interface practically do not change their shape. Approaching the interface, the ferritic grains begin to stretch along the welding direction, and immediately at the interface they repeat the profile of a wave. A highly deformed layer consisting of elongated grains is significantly localized near the interface: its thickness is only a few tens of micrometers. At high-impact velocities, the interface and vortex zones are surrounded by a layer of fine-grained equiaxed ferrite, the formation of which is associated with extremely high temperatures reached due to significant shear strains.

The dark layer, which is most clearly visible on the sample obtained at an impact velocity of 1000 m/s (Fig. 4.9C and K), is of great interest. It is also observed on a sample obtained at an impact velocity of 735 m/s

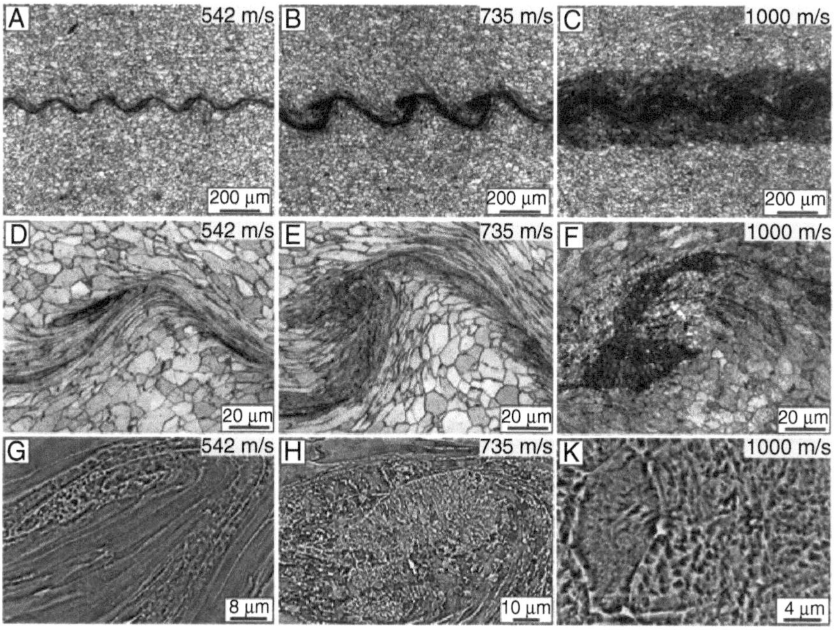

Fig. 4.9 ***Metallographic images of welded samples of low-carbon steel obtained by high-velocity impact of plates in a powder gun. (A–C) Cross sections showing the wave morphology of the interfaces. (D–F) Vortex zones formed on the lateral sides of the waves. (G, H) Scanning electron microscopy images of vortex zones and adjacent material. (K) Scanning electron microscopy image of the dark zone observed in (C).***
From I. A. Bataev, S. Tanaka, Q. Zhou, D. V. Lazurenko, A. M. Jorge Jr., A. A. Bataev, K. Hokamoto, A. Mori, and P. Chen. Towards better understanding of explosive welding by combination of numerical simulation and experimental study. Materials and Design 169: 107649.

(Fig. 4.9B) in the form of a thin band repeating the waveform, but it is almost not observed on a sample obtained at an impact velocity of 542 m/s.

The study of this dark layer using transmission electron microscopy (Fig. 4.10) and comparison of the results with known data on the structure of steels formed under high strain rate compression allows the assumption that it was formed as a result of the $\alpha \rightarrow \varepsilon \rightarrow \alpha$ transformation in iron due to the influence of high pressures. The effect of shock-induced $\alpha \rightarrow \varepsilon \rightarrow \alpha$ transformation on the structure of ferrite grains has been studied in detail by Kheifets et al. [28] and Zel'dovich et al. [29]. In their studies, a significant darkening of ferrite grains during chemical etching was explained by formation of many boundaries during the forward and reverse transitions. The more equiaxed fine-grained ferrite structure, which is closer to the interface, was probably formed as a result of the $\alpha \rightarrow \gamma \rightarrow \alpha$ transformation.

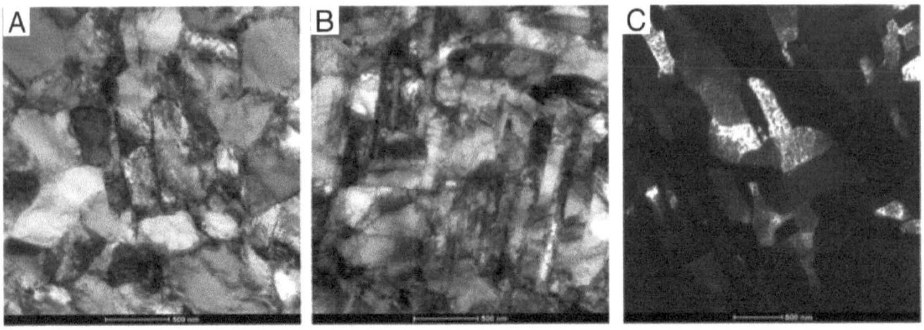

Fig. 4.10 ***Transmission electron microscopy images of structures formed during high-velocity impact welding of low-carbon steel plates. (a) Structures formed as a result of the α → γ → α transformation. (B, C) Bright-field and dark-field images of structures formed in the dark layer as a result of α → ε → α transformation.***

From I. A. Bataev, S. Tanaka, Q. Zhou, D. V. Lazurenko, A. M. Jorge Jr., A. A. Bataev, K. Hokamoto, A. Mori, and P. Chen. Towards better understanding of explosive welding by combination of numerical simulation and experimental study. Materials and Design 169: 107649.

The hypothesis of the ε-phase formation during high-velocity impact welding will be further confirmed using numerical simulation. For this reason, the SPH method was used. To describe the relationship between the thermodynamic parameters in iron during high-velocity impact, a modified Mie–Grüneisen equation of state was used. An empirical Johnson–Cook strength model was used to determine the dependence of flow stress on loading conditions (temperature and strain rate). A more detailed description of the numerical simulation can be found in the work of Bataev et al. [15]. The initial velocities and angles of collision coincided with the velocities and angles measured in the experiments. The results of the simulations and their comparisons with metallographic studies are described in the following.

A comparison of the interface shape observed metallographically and in numerical simulations is shown in Fig. 4.11 [15]. A good correlation between experimental data and simulation results can be noted. The simulation correctly reproduces the wave profile of the interlayer boundary, whereas the amplitude and wavelength coincide well with those observed in the experiments. In addition, the simulation reproduces well the location and size of the melt zones. Therefore, large molten volumes are located on the lateral sides of the waves, whereas on the peaks and depressions the interface is almost free of melt.

Returning to the question of α → ε → α transformation during high-velocity impact welding of steel plates, the following can be noted.

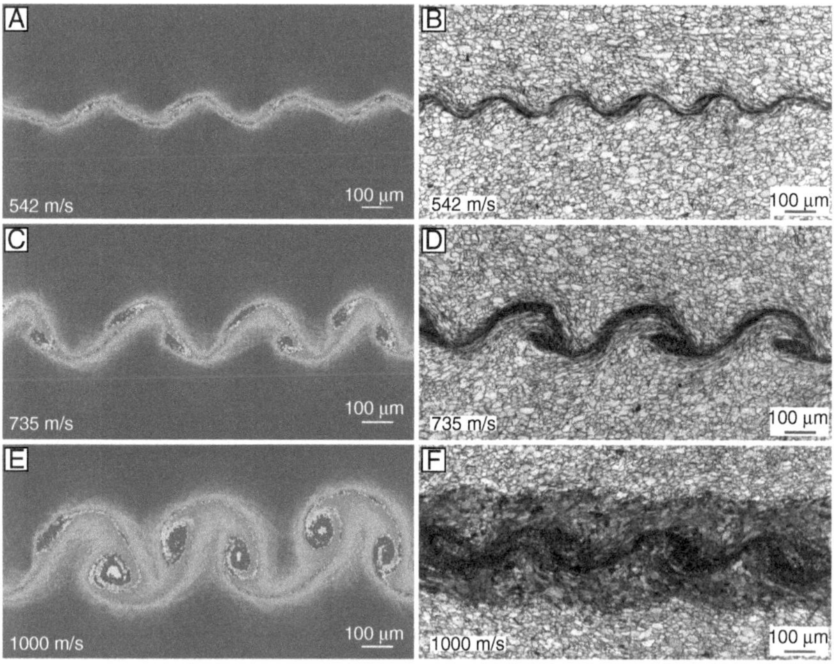

Fig. 4.11 ***Shapes of interfaces observed in the smooth particle hydrodynamics simulation (A, C, E) and in experimental studies (B, D, F). Simulations and experiments were carried out at the same impact velocities. The impact angle in all cases was 15 degrees.*** *From I. A. Bataev, S. Tanaka, Q. Zhou, D. V. Lazurenko, A. M. Jorge Jr., A. A. Bataev, K. Hokamoto, A. Mori, and P. Chen. Towards better understanding of explosive welding by combination of numerical simulation and experimental study. Materials and Design 169: 107649.*

According to the P-T phase diagram of iron [30], at low temperatures the ε-phase is formed in steel at pressures above 10 to 13 GPa. Fig. 4.12 [15] shows the calculated pressure distribution near the collision point. It can be noted that a zone where the pressure exceeds 10 to 13 GPa is well observed only at impact velocity of 1000 m/s. At impact velocity of 735 and 542 m/s, such pressure is achieved only in an extremely narrow zone located along the interface, which is in good agreement with the results of metallographic studies presented in Fig. 4.12B, D, and F.

It is possible to use data on the pressures and temperatures reached at each point of the computational domain to describe the possible phase state of the iron during high-velocity impact welding. The result of such an analysis for the impact velocity of 1000 m/s is shown in Fig. 4.13 [15]. A good agreement can be noted between the shape of the dark region observed metallographically and the calculated shape of the region of

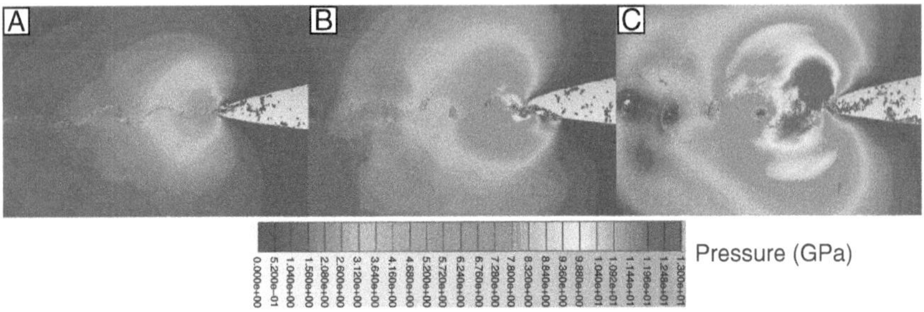

Fig. 4.12 ***The calculated pressure distribution near the collision point during high-velocity impact welding of steel plates at an angle of 15 degrees: 542 m/s (A), 735 m/s (B), and 1000 m/s (C).***

From I. A. Bataev, S. Tanaka, Q. Zhou, D. V. Lazurenko, A. M. Jorge Jr., A. A. Bataev, K. Hokamoto, A. Mori, and P. Chen. Towards better understanding of explosive welding by combination of numerical simulation and experimental study. Materials and Design 169: 107649.

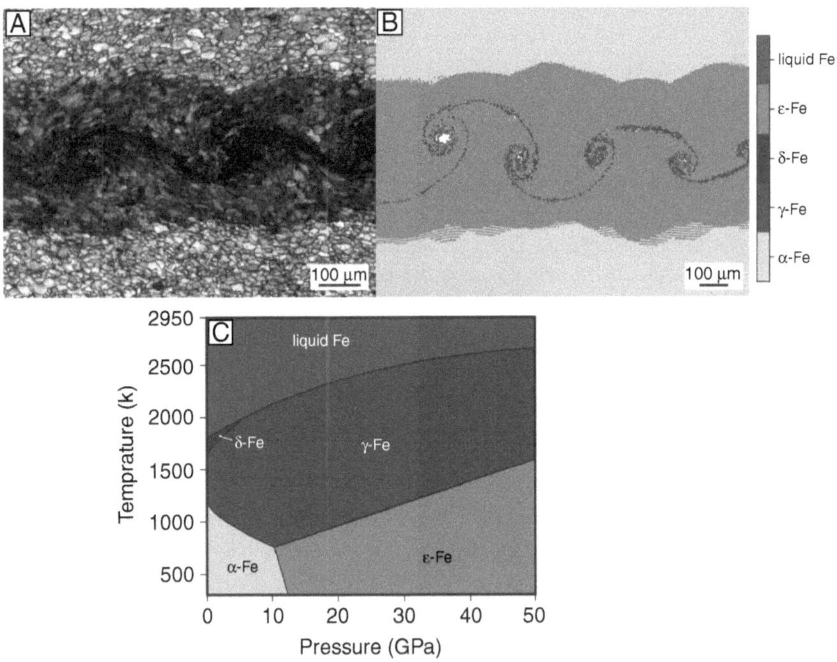

Fig. 4.13 ***Comparison of the dark layer observed by optical microscopy (A) with the layer of ε-phase estimated using smooth particle hydrodynamics numerical simulation (B), and (C) showing the phase diagram of iron in the P-T coordinates used to construct the map in (B).***

From I. A. Bataev, S. Tanaka, Q. Zhou, D. V. Lazurenko, A. M. Jorge Jr., A. A. Bataev, K. Hokamoto, A. Mori, and P. Chen. Towards better understanding of explosive welding by combination of numerical simulation and experimental study. Materials and Design 169: 107649.

ε-phase existence. A thin layer of the γ-phase and molten iron decorates the interface, whereas the vortex zones consist of a swirling mixture of liquid iron and the γ phase. The latter fact is especially interesting, as it helps explain why the materials in vortexes sometimes do not completely melt (due to the increase in pressure, the melting temperature of steel significantly increases).

Based on a combination of metallographic analysis and numerical simulation, it can be concluded that $\alpha \rightarrow \varepsilon \rightarrow \alpha$ transformation is possible in steels under certain conditions of high-velocity impact welding. Despite a significant number of studies devoted to explosive welding of steels, the issue of phase transitions due to pressure increase is rarely considered. At the same time, such transformations can have a significant effect on the mechanical properties of the joint. For instance, in the work of Bataev et al. [15], it is shown that the hardness of the dark layer significantly exceeds the hardness of the surrounding material.

Thus, to analyze the structure of the material subjected to high-velocity oblique impact, three factors must be taken into account: (1) strain hardening due to high shear deformations near the interface, (2) thermal softening due to heating associated with high strain rate deformation, and (3) possible polymorphic transformations near the interface due to an increase in both temperature and pressure. The latter is especially relevant for steels and titanium alloys with various polymorphic modifications at high temperatures and at high pressures.

4.5 Conclusions

Experimentation using a powder gun is a convenient way to analyze the phenomena occurring during explosive welding and other related phenomena. The use of ultra-high-speed cameras allows one to visualize the collision process, accurately measure the angle and velocity of the impact, and observe the formation of the jet. It is convenient to use the SPH method to simulate the processes occurring during the collision of the plates. It allows one to reproduce the formation of waves, vortices, and jets, and can be used to select welding regimes by constructing the welding windows. For a more complete understanding of materials' behavior occurring during high-velocity collisions, it is advisable to use an integrated approach based on observation of the collision process using high-speed cameras, multiscale numerical simulation, and metallographic analysis of the samples obtained.

References

[1] A. Mori, S. Tanaka, K. Hokamoto, Optical observation of metal jet generated by high speed inclined collision, *In Proceedings of SPIE—the 31st International Congress on High-Speed Imaging and Photonics,* Osaka, Japan 2016 November 7–10 (2016) 103281Q-1–103281Q-4.

[2] M.A. Meyers, Dynamic Behavior of Materials, John Wiley & Sons, New York, NY, 1994.

[3] S.K. Godunov, A.A. Deribas, A.V. Zabrodin, N.S. Kozin, Hydrodynamic effects in colliding solids, Journal of Computational Physics 5 (3) (1970) 517–539.

[4] A.A. Deribas, Physics of Explosive Hardening and Welding, [In Russian.] Nauka, Novosibirsk, Russia, 1980.

[5] A.A.A. Mousavi, S.J. Burley, S.T.S. Al-Hassani, W.B. Brown, Simulation of explosive welding with ANFO mixtures, Propellants, Explosives, Pyrotechnics 29 (3) (2004) 188–196.

[6] F. Grignon, D. Benson, K. S.Vechio, and M. A. Meyers. Explosive welding of aluminum to aluminum: analysis, computations and experiments. International Journal of Impact Engineering 30 1333–1351.

[7] A.A.A. Mousavi, S.T.S. Al-Hassani, Numerical and experimental studies of the mechanism of the wavy interface formations in explosive/impact welding, Journal of the Mechanics and Physics of Solids 53 (11) (2005) 2501–2528.

[8] I.A. Bataev, K. Hokamoto, H. Keno, A.A. Bataev, I.A. Balagansky, A.V. Vinogradov, Metallic glass formation at the interface of explosively welded Nb and stainless steel, Metals and Materials International 21 (4) (2015) 713–718.

[9] K. Tanaka, Numerical studies of explosive welding by SPH, Materials Science Forum 566 (2008) 61–64.

[10] X.J. Li, F. Mo, X.H. Wang, B. Wang, K.X. Liu, Numerical study on mechanism of explosive welding, Science and Technology of Welding and Joining 17 (1) (2012) 36–41.

[11] A.S. Bahrani, T.J. Black, B. Crossland, The mechanics of wave formation in explosive welding, Proceedings of the Royal Society A: Mathematical and Physical Sciences 296 (1445) (1967) 123–136.

[12] S. Kakizaki, M. Watanabe, S. Kumai, Simulation and experimental analysis of metal jet emission and weld interface morphology in impact welding, Materials Transactions 52 (5) (2011) 1003–1008.

[13] Q. Zhou, J. Feng, P. Chen, Numerical and experimental studies on the explosive welding of tungsten foil to copper, Materials 10 (9) (2017) 984.

[14] Y. Mahmood, K. Dai, P. Chen, Q. Zhou, A.A. Bhatti, A. Arab, Experimental and numerical study on microstructure and mechanical properties of Ti-6Al-4V/Al-1060 explosive welding, Metals 9 (11) (2019) 1189.

[15] I. A. Bataev, S. Tanaka, Q. Zhou, D. V. Lazurenko, A. M. Jorge Jr., A. A. Bataev, K. Hokamoto, A. Mori, and P. Chen. Towards better understanding of explosive welding by combination of numerical simulation and experimental study. Materials and Design 169: 107649.

[16] H. Liang, N. Luo, T. Shen, X. Sun, X. Fan, Y. Cao, Experimental and numerical simulation study of Zr-based BMG/Al composites manufactured by underwater explosive welding, Journal of Materials Research and Technology 9 (2) (2020) 1539–1548.

[17] A. Nassiri, T. Abke, and G. S. Daehn. Investigation of melting phenomena in solid-state welding processes. Scripta Materialia 168: 61--66.

[18] A. Nassiri, S. Zhang, T. Lee, T. Abke, A. Vivek, B. Kinsey, G. Daehn, Numerical investigation of CP-Ti & Cu110 impact welding using smoothed particle hydrodynamics and arbitrary Lagrangian-Eulerian methods, Journal of Manufacturing Processes 28 (2017) 558–564.

[19] Y.Y. Émurlaeva, I.A. Bataev, Q. Zhou, D.V. Lazurenko, I.V. Ivanov, P.A. Riabinkina, S. Tanaka, P. Chen, Welding window: comparison of Deribas' and Wittman's approaches and SPH simulation results, Metals 9 (12) (2019) 1323.
[20] R.H. Wittman, The influence of collision parameters on the strength and microstructure of the explosion welded aluminum alloy, *In Proceedings of the 2nd International Symposium on Use of an Explosion Energy in Manufacturing Metallic Materials*, Marianske Lazne, Czech Republic, October 9 -12 (1973) 153–168 .
[21] V.I. Lysak, S.V. Kuzmin, Explosive Welding, [In Russian.] Mashinostroenie, Moscow, Russia, 2005.
[22] S.P. Kiselev, V.I. Mali, Numerical and experimental modeling of jet formation during a high-velocity oblique impact of metal plates, Combustion, Explosion, and Shock Waves 48 (2) (2012) 214–225.
[23] S.P. Kiselev, Numerical simulation of wave formation in an oblique impact of plates by the method of molecular dynamics, Journal of Applied Mechanics and Technical Physics 53 (6) (2012) 907–917.
[24] A.A Deribas, V.M. Kudinov, F.I. Matveenkov, V.A. Simonov, Simulation of the process of wave formation in explosive welding, Combustion, Explosion, and Shock Waves 4 (1) (1968) 59–63.
[25] T.T. Zhang, W.X. Wang, J. Zhou, X.Q. Cao, R.S. Xie, Y. Wei, Molecular dynamics simulations and experimental investigations of atomic diffusion behavior at bonding interface in an explosively welded Al/Mg alloy composite plate, Acta Metallurgica Sinica (English Letters) 30 (10) (2017) 983–991.
[26] J. Feng, K. Dai, Q. Zhou, J. Xie, R. Yang, I.A. Bataev, P. Chen, Formation of bonding interface in explosive welding—a molecular dynamics approach, Journal of Physics: Condensed Matter 31 (41) (2019) 415403.
[27] V.I. Mali, V.A. Simonov, Some effects appearing on interactions of shock waves with cavities in metals, *In Proceedings of the 2nd International Symposium on Use of an Explosive Energy in Manufacturing Metallic Materials*, Marianske Lazne, Czech Republic, October 9–12 (1973) 83–96.
[28] A.E. Kheifets, V.I. Zel'dovich, N.Y. Frolova, S.M. Dolgikh, K.V. Gaan, E.V. Shorokhov, Phase and structural transformations in a low-carbon steel that occur upon the collapse of a cylindrical shell, Physics of Metals and Metallography 118 (7) (2017) 681–690.
[29] V.I. Zel'dovich, N.Y. Frolova, A.E. Kheifets, S.M. Dolgikh, K.V. Gaan, E.V. Shorokhov, Deformation- and temperature-related processes that occur upon the collapse of a thick cylindrical shell made of steel 20, Physics of Metals and Metallography 116 (3) (2015) 285–292.
[30] A.T. Dinsdale, SGTE data for pure elements, Calphad 15 (4) (1991) 317–425.

CHAPTER 5

Fabrication of cellular materials with explosive welding

Masatoshi Nishi[a], Kazuyuki Hokamoto[b], Matej Vesenjak[c], Zoran Ren[c,d]
[a]National Institute of Technology, Kumamoto College, Kumamoto, Japan
[b]Professor, Institute of Industrial Nanomaterials, Kumamoto University, Kumamoto, Japan
[c]Faculty of Mechanical Engineering, University of Maribor, Maribor, Slovenia
[d]International Research Organization for Advanced Science and Technology, Kumamoto University, Kumamoto, Japan

5.1 Introduction

Functional cellular materials with a large number of pores and a large specific surface area show high energy absorption and heat exchange capacity, as well as good heat insulation and sound absorption characteristics [1,2]. These materials have widespread applications, including those in biomedicine [3]. A lotus-type porous metal is a useful cellular material with unidirectional pores. Shapovalov [4], Nakajima [5], and Kashihara et al. [6] fabricated the Gasar and lotus-type porous metal by the one-directional solidification process in a pressurized gas atmosphere. However, the lotus-type porous metal with only several-millimeter-long pores can be fabricated by this process. The same type of porous structure, where the cross section, including pores, is uniform through the length, was proposed by Hokamoto et al. [7] through a new process involving explosive compaction and welding. A UniPore structure comprises cylindrical metal pipes processed through explosive compaction to consolidate them. This fabrication process can be used to produce a UniPore structure of up to 1 m in length or even more.

Hokamoto et al. and Kashihara et al. fabricated the UniPore structure by using pure industrial copper [6] and pure industrial aluminum [8]. A modified method based on rolling copper foil was also proposed to simplify the fabrication process [9]. Nishi et al. [10] fabricated composite UniPore structures by using stainless steel and copper pipes. Other studies have included computational simulations to analyze the high-rate deformation process [8,11]. Furthermore, the mechanical properties of the copper UniPore structure [12], the applications for thermal capacitor [13] and heat exchangers [14], and the mechanical response under a high-velocity impact [15] were investigated.

Explosion, Shock-Wave and High-Strain-Rate Phenomena of Advanced Materials
DOI: 10.1016/C2019-0-03954-4

This chapter discusses experimental and computational analyses of the fabrication of UniPore structures by using cylindrical explosive compaction and welding assembly. The UniPore structures were mainly fabricated using pure industrial copper and aluminum pipes. The bonding state between the pipe walls was characterized through microstructural observations, and the high-rate fabrication process was analyzed through computational simulations using the ANSYS AUTODYN multiple-solver-type hydrocode. The mechanical behavior of cellular materials is discussed in Chapter 6.

5.2 Experimental setup

5.2.1 Fabrication conditions of the copper UniPore structure

Fig. 5.1 shows a schematic diagram of the experiment. The fabrication assembly consists of the outer copper pipe and the inner copper pipes filled with paraffin to prevent complete compaction. This study used JIS-C1220 standard copper. Three samples with inner pipes having a thickness of 0.5,

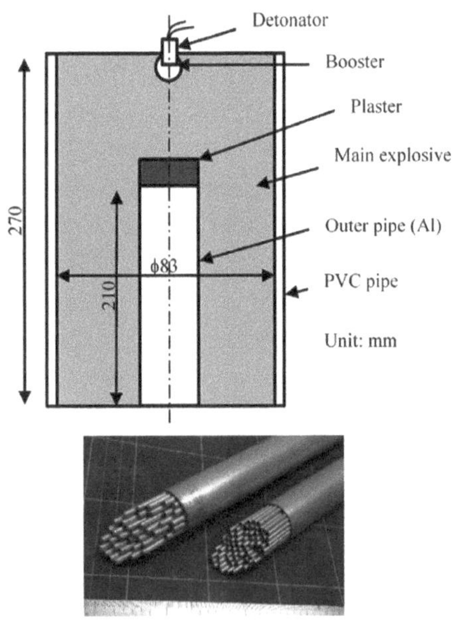

Fig. 5.1 ***Experimental assembly.*** *(Reprinted from K. Hokamoto, K. Shimomiya, M. Nishi, L. Krstulović-Opara, M. Vesenjak, and Z. Ren. 2018. Fabrication of unidirectional porous-structured aluminum through explosive compaction using cylindrical geometry.* Journal of Materials Processing Technology *251: 262–266 with permission from Elsevier).*

Table 5.1 Dimensions of the copper UniPore structure.

	Outer pipe		Inner pipe			
Symbol (porosity)	Diameter (mm)	Length (mm)	Diameter (mm)	Length (mm)	Number	Initial porosity
High	Ø30/Ø27	200	Ø3.0/Ø2.6	200	63	0.65
Medium	Ø30/Ø27	200	Ø3.0/Ø2.4	200	63	0.58
Low	Ø30/Ø27	200	Ø3.0/Ø2.0	200	63	0.45

0.3, and 0.2 mm, corresponding to low, medium, and high porosity, respectively, were fabricated. Table 5.1 lists their dimensions. ANFO-A explosive comprising 750 g of ANFO powder was used as the main explosive [16,17] filling in the gap between the outer pipe and the PVC pipe. The ANFO-A explosive was detonated using 10 g of SEP explosive as a booster. The generated high pressure and detonation gas acted on the outer pipe by accelerating it to high velocity toward the center axis. Upon high-speed collision with inner pipes, the outer and inner pipes were firmly welded tightly together via explosive welding phenomena [18].

5.2.2 Fabrication conditions of the aluminum UniPore structure

The experimental method for fabricating the aluminum UniPore structure [8] is the same as that for the copper structure. As shown in Table 5.2, four samples were fabricated by combining two different types of outer and inner pipes; the dimensions are also listed.

5.3 Computational simulation setup

Since the real-time fabrication process of the UniPore structures is almost impossible to observe, the computational simulations using ANSYS AUTODYN were used to analyze the fabrication process, especially the

Table 5.2 Dimensions of the aluminum UniPore structure.

	Outer pipe		Inner pipe			
Symbol	Diameter (mm)	Length (mm)	Diameter (mm)	Length (mm)	Number	Initial porosity
LL	Ø30/Ø28	200	Ø3.0/Ø2.6	200	68	0.63
LS	Ø32/Ø26	200	Ø3.0/Ø2.6	200	61	0.46
SL	Ø30/Ø28	200	Ø4.0/Ø2.6	200	37	0.21
SS	Ø32/Ø26	200	Ø4.0/Ø2.6	200	32	0.14

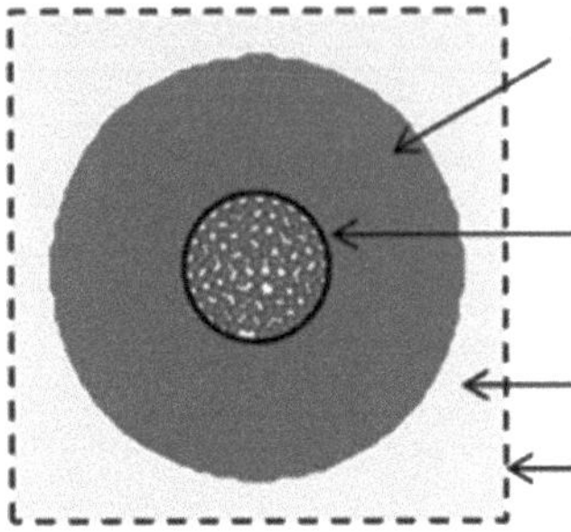

Fig. 5.2 ***Analytical model of the aluminum UniPore structure (LL sample).*** *(Reprinted from K. Hokamoto, K. Shimomiya, M. Nishi, L. Krstulović-Opara, M. Vesenjak, and Z. Ren. 2018. Fabrication of unidirectional porous-structured aluminum through explosive compaction using cylindrical geometry.* Journal of Materials Processing Technology *251: 262–266 with permission from Elsevier).*

collision velocity between pipes and pipe failures [11]. Computational simulations were conducted for the three copper UniPore structures (Subsection E2-1) and four aluminum UniPore structures (Subsection E2-2). The UniPore assembly with a transverse cross-sectional view was modeled using a planar 2D model. Fig. 5.2 shows a computational simulation model of the aluminum UniPore structure (LL sample). The inner pipes filled with paraffin and outer pipes were modeled using a Lagrangian formulation, and interactive boundary conditions implemented in ANSYS AUTODYN were applied between these elements. The multiple-material Eulerian solver was used to model the ANFO-A explosive and the void domain. The Euler–Lagrange interaction effect was applied to the outer surfaces of the outer pipe.

The Mie–Grüneisen equation of state [19] was applied for all pipes and paraffin, and the Johnson–Cook constitutive law [20] was set for all pipes. The Mie–Grüneisen equation of state is expressed in Eq. (1) with the parameters in Table 5.3:

$$P = p_H + \Gamma \rho \left(e - e_H\right) \qquad ((5.1)$$

$$p_H = \frac{\rho_0 \, c_0^2 \, \mu \left(1 + \mu\right)}{\left[1 - \left(s - 1\right) \mu\right]^2}$$

$$e_H = \frac{1}{2} \frac{p_H}{\rho_0} \left(\frac{\mu}{1 + \mu}\right)$$

Table 5.3 Parameters of the Mie–Grüneisen equation of state based on shock Hugoniot (index "0" denotes the initial variable).

	Reference density(ρ_0/kg • m^{-3})	Grüneisen coefficient (Γ/-)	Velocity of sound (c_0/m•s^{-1})	Material constant (*s*/-)
Copper	8900	2.0	3958	1.497
Aluminum	2707	1.9	5386	1.339
Paraffin	918	1.18	2908	1.56

The material models for copper, aluminum, and paraffin were obtained from the ANSYS AUTODYN material library.

For the ANFO-A explosive, the ideal gas equation that considers only the high-pressure gas was applied in these computational simulations. The gas equation and its material parameters for the ANFO-A explosive were cited in the work of Fujita et al. [21].

5.4 Results and discussion

This section discusses the simulation results of the deformation process. The final deformed shape predicted by computational simulations was analyzed and compared with the experimental results. The collision velocity between pipes and damaged pipes together with microstructural analysis of the samples are also discussed.

5.4.1 Deformation process

Fig. 5.3 shows the simulated explosive welding process of the aluminum UniPore structure (LL sample) upon circumferential explosive detonation. The collision velocity between the pipes is an essential parameter in explosive welding, and it must exceed 200 m/s for tight bonding of metals [18].

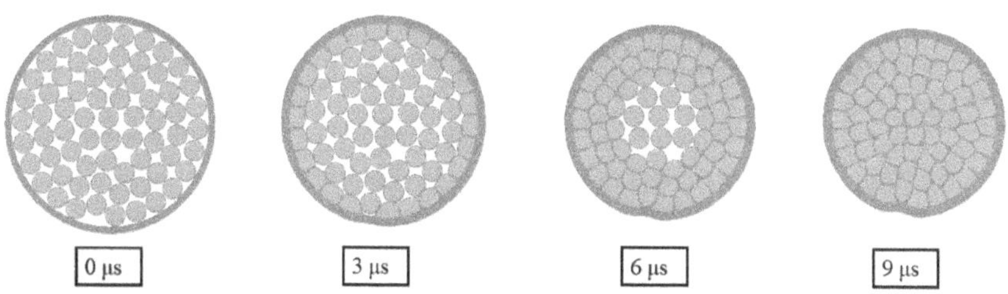

Fig. 5.3 ***Deformation process predicted through computational simulation (explosive is not displayed).***

5.4.2 Final deformed shape of samples with different porosities

5.4.2.1 Copper UniPore structure

Three samples were successfully fabricated by explosive compaction, as shown in Fig. 5.4. The inner and outer pipes were joined without a gap in the cross-sectional area. Fig. 5.5 shows the final deformed shapes of the three types of samples obtained by computational simulations. For all

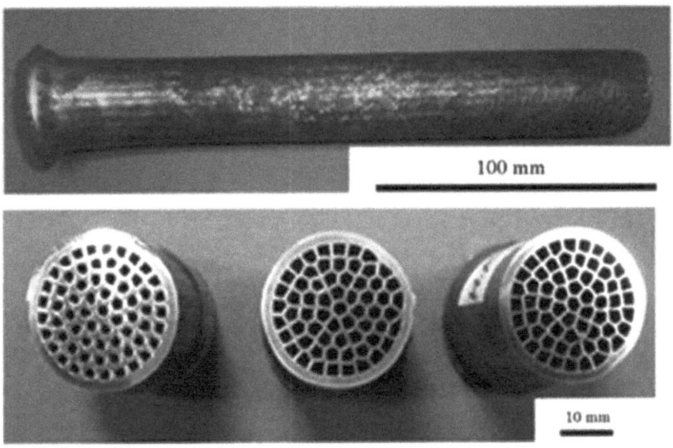

Fig. 5.4 ***Appearance of recovered low-, medium-, and high-porosity samples (from* left *to* right*) in experimental investigations.*** *(Reprinted from K. Hokamoto, M. Vesenjak, and Z. Ren. 2014. Fabrication of cylindrical uni-directional porous metal with explosive compaction.* Materials Letters *137: 323–327 with permission from Elsevier).*

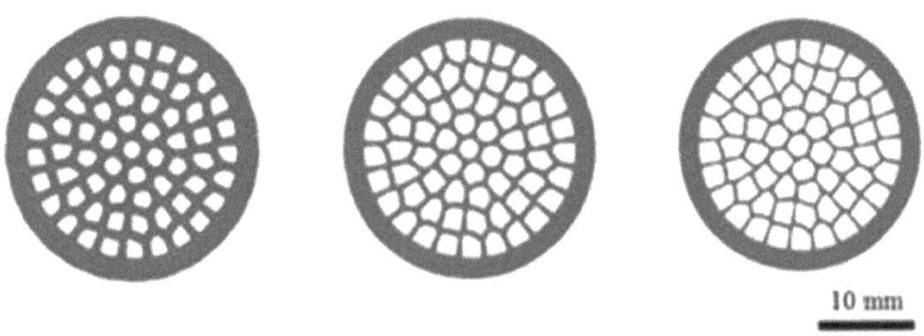

Fig. 5.5 ***Final deformed shape of low-, medium-, and high-porosity samples (from* left *to* right*) as predicted by computational simulations.*** *(Reprinted from M. Nishi, M. Oshita, M. Ulbin, M. Vesenjak, Z. Ren, and K. Hokamoto. 2018. Computational analysis of the explosive compaction fabrication process of cylindrical uni-directional porous copper.* Metals and Materials International *24: 1143–1148 with permission from Springer).*

samples, the computational results of the final deformed shape were similar to the experimental results. A comparison of both results indicates that the computational simulations can be used to predict the final deformed shape that would be produced in the experiment.

5.4.2.2 Aluminum UniPore structure

Fig. 5.6 shows the recovered samples (LL, LS, SL, and SS from top to bottom) and their cross sections (LL, LS, SL, and SS from left to right). The fracture of the outer pipe can be observed in the LL sample with thin outer and inner pipes, whereas the other specimens were recovered without any cracks. This indicates that an appropriate thickness of the pipe wall should be selected. The cross-sectional views in Fig. 5.6 suggest that many pores have similar areas and elongated holes are formed parallel to the vertical axis. Fig. 5.7 shows the final deformation shapes of the samples as calculated by the computational analysis. The results of the cross-sectional views are

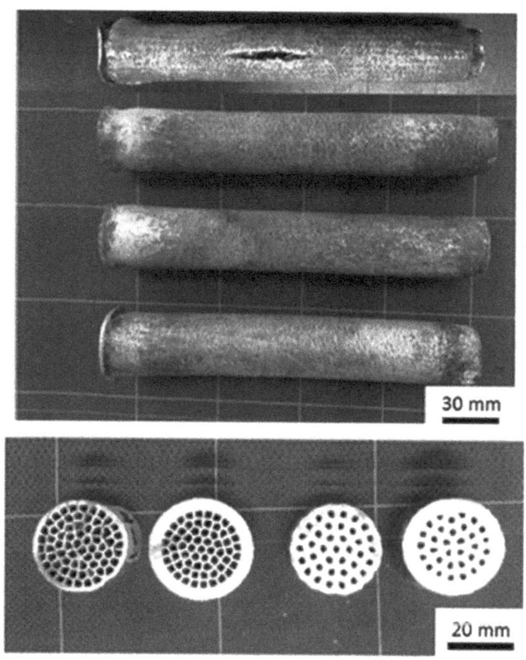

Fig. 5.6 ***Appearance of samples recovered.*** *(Reprinted from K. Hokamoto, K. Shimomiya, M. Nishi, L. Krstulović-Opara, M. Vesenjak, and Z. Ren. 2018. Fabrication of unidirectional porous-structured aluminum through explosive compaction using cylindrical geometry.* Journal of Materials Processing Technology *251: 262–266 with permission from Elsevier).*

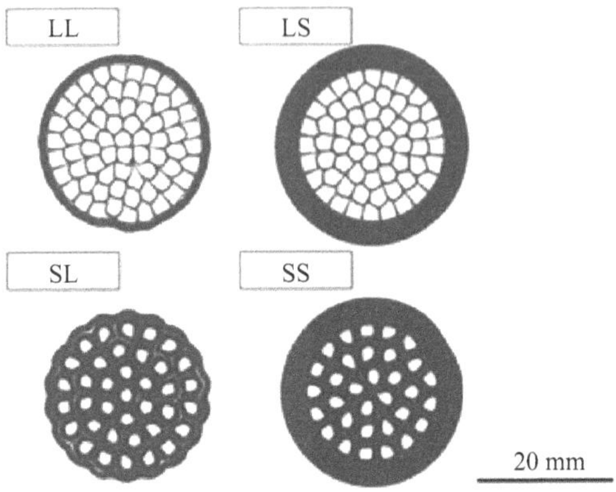

Fig. 5.7 ***Final deformed shape for four samples (computational simulations).*** *(Reprinted from K. Hokamoto, K. Shimomiya, M. Nishi, L. Krstulović-Opara, M. Vesenjak, and Z. Ren. 2018. Fabrication of unidirectional porous-structured aluminum through explosive compaction using cylindrical geometry.* Journal of Materials Processing Technology *251: 262–266 with permission from Elsevier).*

similar to the experimental results shown in Fig. 5.6, as in the case of copper samples. The computational analysis was 2D; therefore, the damage of the outer pipe as in the LL sample could not be replicated.

5.4.3 Microstructure of recovered samples

5.4.3.1 Copper UniPore structure

This section discusses the microstructure of the high-porosity sample. The bottom part of Fig. 5.8 shows a trace of Vickers penetrators under 10 gf (9.8 mN) load; the Vickers hardness of typical annealed copper is 60 HV. The upper right and bottom of Fig. 5.8 show the melted area at the triple collision point of inner pipes; the measured Vickers hardness at this point was relatively low, being 44 to 61 HV. Therefore, shrinkage cavities or other defects may occur in a limited region. In contrast, the pipe hardness in other areas generally increased because work hardening occurred via the collision of inner pipes with hardness in the range of 130 to 150 HV. The average hardness of the as-received pipes was 120 HV.

Fig. 5.9 shows the pressure contour in the sample during the fabrication process. High pressure occurs at the point where the three pipes collide. This point is the same as the melted area shown in Fig. 5.8; it is suggested that high pressure and high temperature occur at this point due to compression toward the axial center.

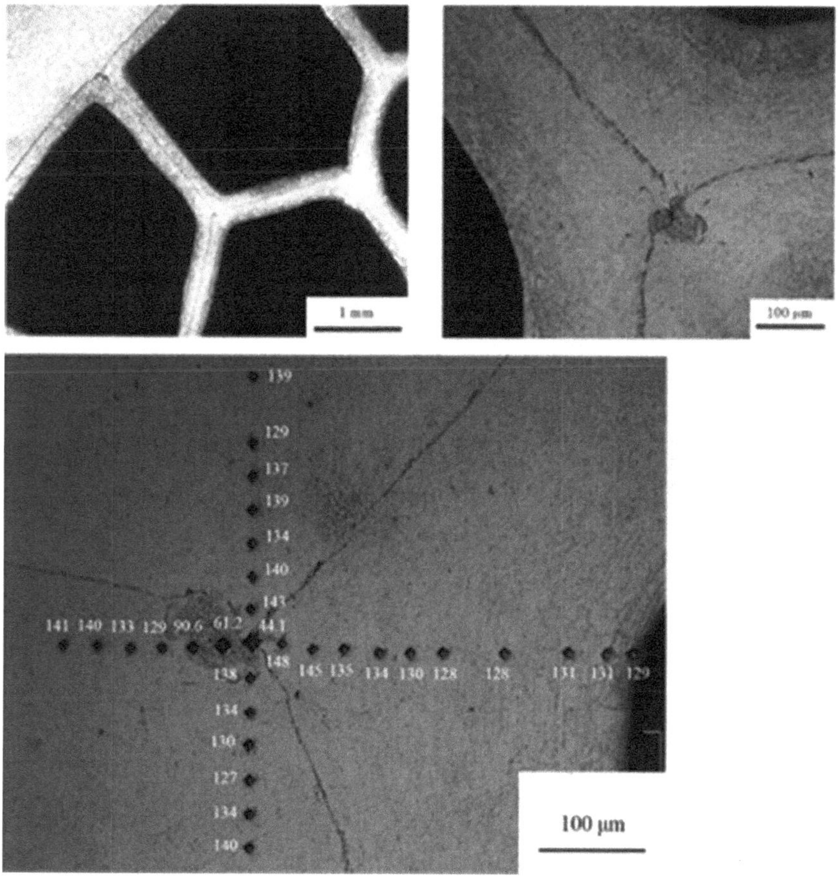

Fig. 5.8 ***Microstructure of cross section and trace of Vickers penetrator.*** *(Reprinted from K. Hokamoto, M. Vesenjak, and Z. Ren. 2014. Fabrication of cylindrical uni-directional porous metal with explosive compaction.* Materials Letters *137: 323–327 with permission from Elsevier).*

5.4.3.2 Aluminum UniPore structure

Fig. 5.10 shows an enlarged sectional view of the experimental results for the LL sample. Fractures of inner pipes were occasionally observed. Fig. 5.11 shows the final deformation shape (LL sample) obtained by computational analysis. A part of the inner pipe is massively deformed during fabrication, and this is considered to be the cause of pipe failure and corresponds with the experimental results shown in Fig. 5.10. Such failure is likely to occur when the pipe is thin and there is a large gap in the initial placement between pipes [8]. Therefore, the experimental setup must be carefully prepared to avoid large deformation and failure of the inner

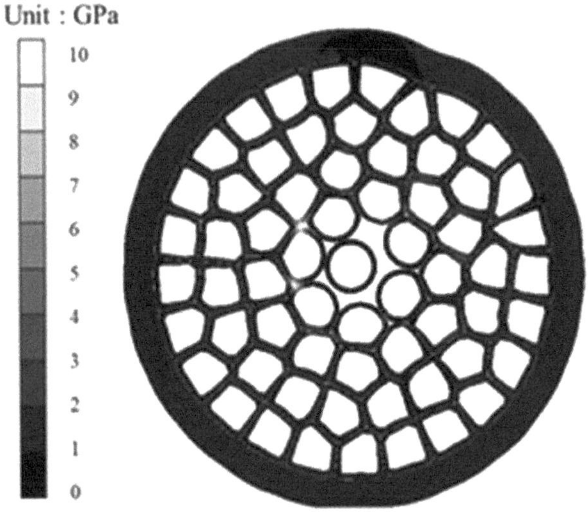

Fig. 5.9 ***Pressure contours during fabrication of the UniPore structure.***

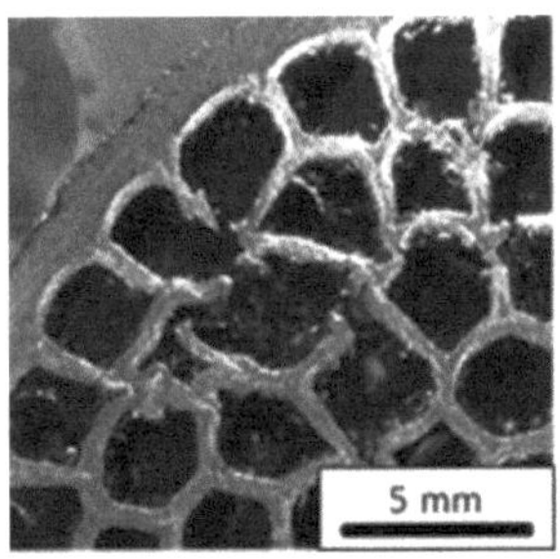

Fig. 5.10 ***Microstructure of transverse cross section of the LL sample.*** *(Reprinted from K. Hokamoto, K. Shimomiya, M. Nishi, L. Krstulović-Opara, M. Vesenjak, and Z. Ren. 2018. Fabrication of unidirectional porous-structured aluminum through explosive compaction using cylindrical geometry.* Journal of Materials Processing Technology *251: 262–266 with permission from Elsevier).*

pipes. In the other samples (LS, SL, and SS), the inner and outer pipes were welded successfully without any fracture and gaps [8].

5.4.4 Pipe velocity during fabrication

5.4.4.1 Copper UniPore structure

Fig. 5.12 shows the history of the absolute velocity of the outer pipe of the medium-porosity sample. The velocity history for initial distances of 0.2 and 0.8 mm between the outer and inner pipes is indicated by the dotted and solid

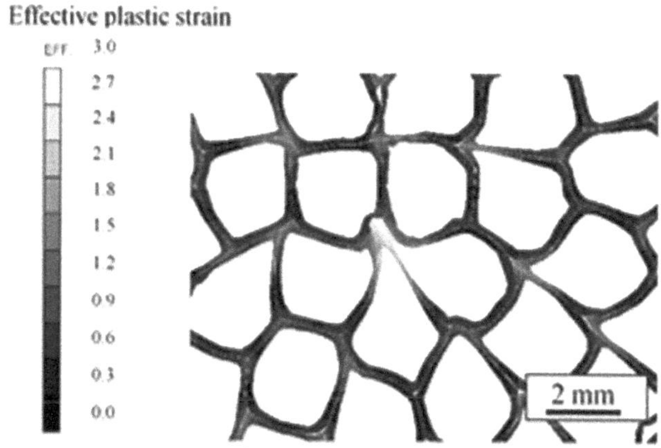

Fig. 5.11 ***Enlarged view of the LL sample.*** *(Reprinted from K. Hokamoto, K. Shimomiya, M. Nishi, L. Krstulović-Opara, M. Vesenjak, and Z. Ren. 2018. Fabrication of unidirectional porous-structured aluminum through explosive compaction using cylindrical geometry.* Journal of Materials Processing Technology *251: 262–266 with permission from Elsevier).*

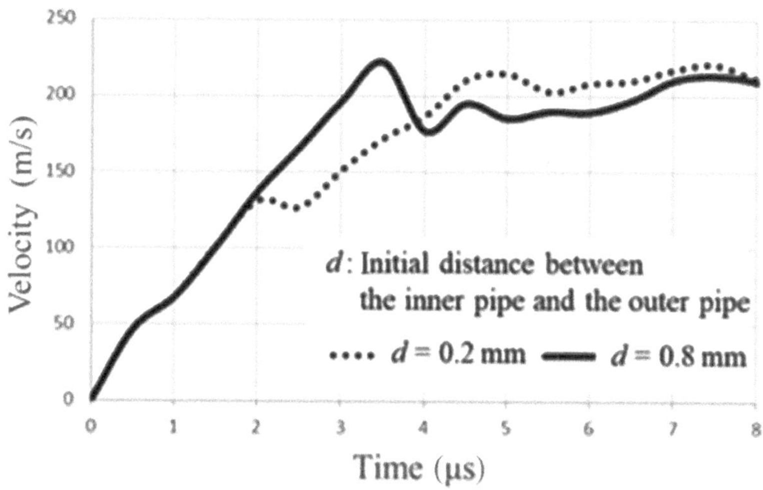

Fig. 5.12 ***Initial velocity of the outer pipe.*** *(Reprinted from M. Nishi, M. Oshita, M. Ulbin, M. Vesenjak, Z. Ren, and K. Hokamoto. 2018. Computational analysis of the explosive compaction fabrication process of cylindrical uni-directional porous copper.* Metals and Materials International *24: 1143–1148 with permission from Springer).*

lines, respectively. The pipes are observed to collide at a submarginal velocity for explosive welding [18]. The velocity of the outer pipe remained almost the same in the low- and high-porosity cases. The bonding condition between

Fig. 5.13 ***Transversal compressive testing.*** *Reprinted from M. Vesenjak, K. Hokamoto, M. Sakamoto, T. Nishi, L. Krstulović-Opara, and Z. Ren. 2016. Mechanical and microstructural analysis of unidirectional porous (UniPore) copper.* Materials and Design *90: 867–880 with permission from Elsevier.*

pipes may be insufficient in some cases due to insufficient collision velocity. The outer and inner pipes have been reported to peel off under compressive loading, as shown in Fig. 5.13 [12]. To achieve sufficient bonding between the pipes and to meet the requirements of bonding using explosive welding, it is necessary to include a gap between the outer and inner pipe at initial placement for an efficient acceleration of the outer pipe to achieve higher collision velocities.

5.4.4.2 Aluminum UniPore structure

Fig. 5.14 shows the change in the absolute velocity V of the outer pipe for the LL sample in time. A sufficient collision velocity for explosive welding is confirmed for the aluminum UniPore structure even with initial pipe placement similar to that of the copper structure. This is due to the difference in the material strength of the base material of the UniPore structure.

5.5 Studies on UniPore structures

The inner pipes must be filled completely with paraffin to prevent their collapse during fabrication (Fig. 5.15). Therefore, a simplified manufacturing method has been proposed, in which a rolled aluminum foil attached with a square PMMA bar is inserted into the outer pipe [9] (see Fig. 5.15). The composite UniPore structures shown in Fig. 5.16 are fabricated using stainless steel and copper pipes [10]. This structure has the anticorrosion effect of stainless steel and excellent heat transfer performance of copper.

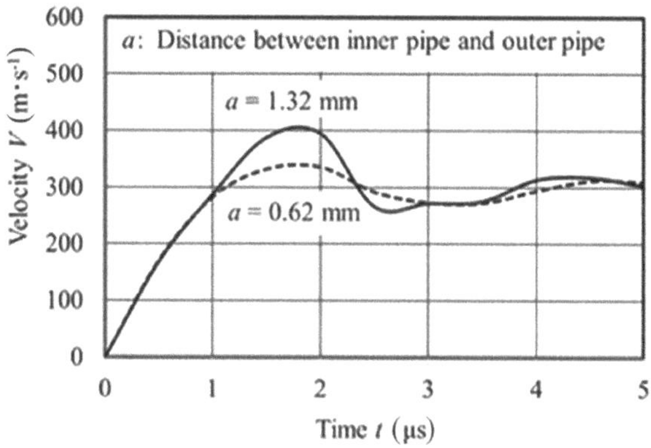

Fig. 5.14 ***Velocity of the outer pipe at the initial stage.*** *(Reprinted from K. Hokamoto, K. Shimomiya, M. Nishi, L. Krstulović-Opara, M. Vesenjak, and Z. Ren. 2018. Fabrication of unidirectional porous-structured aluminum through explosive compaction using cylindrical geometry.* Journal of Materials Processing Technology *251: 262–266 with permission from Elsevier).*

Fig. 5.15 ***Rolled foil inserted into the outer pipe*** **(left)** ***and experimental assembly*** **(right)**. *(Reprinted from M. Vesenjak, K. Hokamoto, S. Matsumoto, Y. Marumo, and Z. Ren. 2016. Unidirectional porous metal fabricated by rolling of copper sheet and explosive compaction. Materials Letters 170: 39–43 with permission from Elsevier).*

Fig. 5.16 ***Composite UniPore structure.*** *(Reprinted from M. Nishi, S. Tanaka, M. Vesenjak, Z. Ren, and K. Hokamoto. 2020. Fabrication of composite unidirectional cellular metals by using explosive compaction.* Metals (Basel) *10(2): 193).*

Studies have investigated the application of UniPore structures as thermal capacitors [13], heat exchangers [14], and mechanical response measurement under high-velocity impact [15]. In particular, Sato et al. [14] measured high heat transfer performance in a gas flow by using the copper UniPore structure, although they had to make efforts to reduce pressure loss. When achieving a high-velocity impact at approximately 400 m/s by using a powder gun, high pressure with intense jetting spread outside the pore during void (pore) collapse [14]. Numerical simulations indicate that inducing a high pressure exceeding 25 GPa could enable the synthesis of new materials [14].

5.6 Conclusions

This chapter discusses the manufacturing method to fabricate UniPore structures, a type of cellular metal, by using pure industrial copper and pure industrial aluminum as base materials. Here, the manufacturing method is investigated in detail through experiments and computational analyses. The following results were obtained through the investigation:

1. Microstructure observations confirmed that pipes with a UniPore structure show good bonding due to the high pressure generated by the explosive. In addition, partial melting occurs at the triple point where the inner pipes collide.
2. The experimental results are supported by computational analyses of the forming process during compaction, collision velocity of the outer pipe, and predicted final deformation shape.
3. Pipe wall fracture occurred in samples with low pipe thickness. Therefore, an appropriate inner wall thickness and initial placement of an inner pipe without a large gap should be selected.

References

[1] J. Banhart, Manufacture, characterisation and application of cellular metals and metal foams, Programs in Materials Science 46 (6) (2001) 559–632.

[2] M. Ashby, A. Evans, N. Fleck, L. Gibson, J. Hutchinson, H. Wadley, F. Delale, Metal foams: a design guide, Applied Mechanics Review 54 (6) (2001) B015–B106.

[3] H. Attar, S. Ehtemam-Haghighi, N. Soro, D. Kent, M.S. Dargusch, Additive manufacturing of low-cost porous titanium-based composites for biomedical applications: advantages, challenges and opinion for future development, Journal of Alloys and Compounds 827 (2020) 154263.

[4] V. Shapovalov, Porous metals, MRS Bulletin 19 (4) (1994) 24–28.

[5] H. Nakajima, Fabrication, properties and application of porous metals with directional pores, Progress in Materials Science 52 (7) (2007) 1091–1173.

[6] M. Kashihara, S.K. Hyun, H. Yonetani, T. Kobi, H. Nakajima, Fabrication of lotus-type porous carbon steel by unidirectional solidification in nitrogen atmosphere, Scripta Materialia 54 (4) (2006) 509–512.
[7] K. Hokamoto, M. Vesenjak, Z. Ren, Fabrication of cylindrical uni-directional porous metal with explosive compaction, Materials Letters 137 (2014) 323–327.
[8] K. Hokamoto, K. Shimomiya, M. Nishi, L. Krstulović-Opara, M. Vesenjak, Z. Ren, Fabrication of unidirectional porous-structured aluminum through explosive compaction using cylindrical geometry, Journal of Materials Processing Technology 251 (2018) 262–266.
[9] M. Vesenjak, K. Hokamoto, S. Matsumoto, Y. Marumo, Z. Ren, Uni-directional porous metal fabricated by rolling of copper sheet and explosive compaction, Materials Letters 170 (2016) 39–43.
[10] M. Nishi, S. Tanaka, M. Vesenjak, Z. Ren, K. Hokamoto, Fabrication of composite unidirectional cellular metals by using explosive compaction, Metals (Basel) 10 (2) (2020) 193.
[11] M. Nishi, M. Oshita, M. Ulbin, M. Vesenjak, Z. Ren, K. Hokamoto, Computational analysis of the explosive compaction fabrication process of cylindrical uni-directional porous copper, Metals and Materials International 24 (2018) 1143–1148.
[12] M. Vesenjak, K. Hokamoto, M. Sakamoto, T. Nishi, L. Krstulović-Opara, Z. Ren, Mechanical and microstructural analysis of unidirectional porous (UniPore) copper, Materials and Design 90 (2016) 867–880.
[13] T. Fiedler, M. Borovinšek, K. Hokamoto, M. Vesenjak, High-performance thermal capacitors made by explosion forming, International Journal of Heat and Mass Transfer 83 (2015) 366–371.
[14] Y. Sato, K. Yuki, Y. Abe, R. Kibushi, N. Unno, K. Hokamoto, S. Tanaka, T. Tomimura, Heat transfer characteristics of a gas flow in uni-directional porous copper pipes, Proceedings of the 16th International Heat Transfer Conference, 2020 8189–8193 August 10–15.
[15] M. Nishi, S. Tanaka, M. Vesenjak, Z. Ren, K. Hokamoto, Experimental and computational analysis of the uni-directional porous (UniPore) copper mechanical response at high-velocity impact, International Journal of Impact Engineering 136 (2020) 103409.
[16] I.A. Bataev, K. Hokamoto, H. Keno, A.A. Bataev, I.A. Balagansky, A.V. Vinogradov, Metallic glass formation at the interface of explosively welded Nb and stainless steel, Metals and Materials International 21 (2015) 713–718.
[17] A.M. Satyanarayan, M. Nishi, K. Hokamoto, Underwater shock wave weldability window for Sn-Cu plates, Journal of Materials Processing Technology 267 (2019) 152–158.
[18] C. Bernard, Explosive Welding of Metals and Its Application, Clarendon Press, Oxford, UK, 1982.
[19] R.G. McQueen, S.P. Marsh, J.W. Taylor, J.N. Frits, W.J. Carter, The equation of state of solids from shock wave studies, in: R. Kinslow (Ed.), High-Velocity Impact Phenomena, Academic Press, San Diego, CA, 1970, pp. 249–419.
[20] G.R. Johnson, W.H. Cook, A constitutive model and data from metals subjected to large strains, high strain rates and high temperatures, Proceedings of the 7th International Symposium on Ballistics, 1983 541–547 April 19–21.
[21] M. Fujita, A. Chiba, I. Fukuda, H. Osaka, F. Baba, T. Manabe, An analysis of multilayered explosive bonding process. [In Japanese.], Kyogyo-Kayaku 48 (3) (1987) 176–182.

CHAPTER 6

Mechanical behavior of cellular materials—from quasistatic to high strain rate impact response

Nejc Novak[a], Matej Vesenjak[a], Masatoshi Nishi[b], Shigeru Tanaka[c], Kazuyuki Hokamoto[d], Zoran Ren[e]
[a]Faculty of Mechanical Engineering, University of Maribor, Maribor, Slovenia
[b]National Institute of Technology, Kumamoto College, Kumamoto, Japan
[c]Kumamoto University, Japan
[d]Professor, Institute of Industrial Nanomaterials, Kumamoto University, Kumamoto, Japan
[e]Faculty of Mechanical Engineering, University of Maribor, Maribor, Slovenia; International Research Organization for Advanced Science and Technology, Kumamoto University, Kumamoto, Japan

6.1 Introduction

The lack of resources, rising cost of raw materials, and more stringent policies on sustainable development have resulted in an ever-increasing demand for the use of lighter and multifunctional materials with improved properties in engineering and other applications. One of the promising material types for such applications are cellular materials, such as materials with internal cellular structure either on nano-, micro-, meso-, or macroscale. Combining different base materials with purposely designed internal cellular structure of different morphologies enables us to achieve a unique and attractive combination of their mechanical and thermal properties. These can be specifically tailored by careful design of the material cellular structure.

Metal foams and different types of cellular structures have been comprehensively investigated under various aspects and considered for use in modern applications in technology, medicine, fashion, and other areas [1,2]. Due to their attractive combination of lightweight, high-energy absorption capacity, damping, and thermal insulation, they are one of the most promising building materials for future products [3]. Cellular materials have some advantageous mechanical and thermal properties compared to solid materials [1], such as low density, high acoustic insulation and damping, better thermal management (for use in heat exchangers, flame arresters, heat shields), high energy absorption capacity (for crash absorbers), durability under dynamic load and fatigue, and filters. The manufacturing methods of cellular materials are described in detail in the work of Ashby et al. [2] and Banhart [4].

Explosion, Shock-Wave and High-Strain-Rate Phenomena of Advanced Materials
DOI: 10.1016/C2019-0-03954-4

The foams can be divided into two main groups considering their cell morphology: open- and closed-cell foams [5]. The open-cell foams are used for functional applications (e.g., filtration, catalysis, heat transfer, and biomedical applications). In contrast, the closed-cell foams are mostly used for structural applications, typically as weight-saving and shock-absorbing structures in vehicles.

A classification system based on various levels of geometric control and related production methods was recently introduced by Lehmus et al. [3]:

- *Stochastic foams*
  - Melt foaming using blowing agents, gas introduction (uncontrolled)
  - Powder metallurgy using blowing agents
- *Partially ordered foams: enhanced structural control on the pore level*
  - Melt foaming using the controlled gas introduction
  - Melt foaming using solidification control
  - Placeholder methods
  - Precursor/replication methods
  - Advanced pore morphology foams
  - Syntactic foams
  - Placeholder methods with the ordered assembly of placeholders
  - Syntactic foams through the ordered assembly of hollow particles
  - Metal fiber structures
- *Ordered foams: ordered assembly of identical structural elements*
  - Kagome-type foams
  - Corrugated sheet metal structures
  - UniPore structures
- *Designed-to-purpose cellular structures*
  - Additive manufacturing approaches.

Some cellular materials are already used in practical applications [6]. However, most have only been experimentally and computationally investigated [7–10] but so far never used in real-life applications due to current high production costs. Most mass production technologies lack in the ability to control the shape, size, and distribution of cell pores during the production process, resulting in certain scattering of mechanical properties. This consequently somewhat limits their applicability. New unidirectional structures such as Lotus [9] and UniPore [10], metallic hollow sphere structures [11,12], and advanced pore morphology elements [11–13] have recently been introduced to overcome this shortcoming. Recent advances in additive manufacturing technologies enable the creation of well-structured cellular structures with constant or spatially graded porosity distribution. Additive manufacturing also makes it possible to fabricate the internal structure of cellular materials that is designed in a way that their deformation leads to nonconventional behavior, such as the negative Poisson ratio (ν) [14]. Materials with a negative Poisson ratio sustain large volumetric change under deformation and exhibit nonlinear change in stiffness. Characteristically, they extend laterally when stretched and shrink when compressed. Such materials are often called auxetic materials [15].

Computational simulations based on the finite element method are a vital tool to evaluate the deformation behavior and mechanical response of various types of cellular materials and their use in potential applications. The computational models can be based on homogenized computational models [16], beam finite element models [17], shell finite element models [18], and volume finite element models [19], depending on the geometry and level of desired precision of geometry description. However, they always need to be validated to achieve a good correlation between computational and experimental results. The computational models are often used to optimize the internal cellular structure to achieve the desired response of the material or component [20]. Advanced computational simulations can also be used for validation and observation of the fabrication procedure [21].

The development process of novel cellular structures is a synergy of the manufacturing process, experimental characterization, and computational simulations. As such, an accurate determination of mechanical properties and prediction of deformation behavior is assured as the necessity for a proper design of the final application of newly developed cellular structures.

6.2 Mechanical characterization of cellular structures

Most tests of cellular structures are carried out under compressive loading conditions where significant energy absorption can be observed due to the unique mechanical response (elastic deformation followed by long plateau stress and densification).

Uniaxial compression tests under quasistatic and dynamic loading conditions for most analyzed samples presented in this chapter were performed with the servo-hydraulic testing machine INSTRON 8801 with a position-controlled crosshead velocity of 0.1 mm/s (quasistatic test) and 284 mm/s (dynamic test). If different, the velocity is given in the description of the experiment. The tests were carried out according to the standard ISO 13314: 2011 [22]. The recorded load-displacement data was converted into engineering stress-strain data using the dimensions of the nondeformed sample. The load plates of the testing machine were lubricated to reduce friction and thus minimize the effects of the bearing/load limit on the specimens.

The deformation can also be tracked by using infrared (IR) thermography. IR thermography, based on the fast cooled middle wave (MW) IR InSb detector camera, detects the heat dissipation during the irreversible plastic deformation process [23]. The acquisition of dynamic processes is of particular interest due to the high frame rates of thermal image detectors—for example, up to 700 Hz for the used InSb detector camera, which operates at cryogenic temperatures of -200 °C (73 K). Such an IR-based approach makes it possible to follow the propagation of plastification front waves, plastification areas, and crack growth. Cooled medium IR wave cameras with image frequencies above 100 Hz allow fast acquisition of deformation patterns due to material heating during the irreversible dynamic deformation. This method has already proven to be suitable for observing deformation patterns in various cellular materials under dynamic loading conditions [7,24]. The method is limited to dynamic load cases due to the dissipation of heat at lower loading velocities. Visualization of the local heat dissipation of heat-conducting metals is not possible for quasistatic loading. IR thermographic monitoring was applied during dynamic compression of the auxetic samples with the cooled mediumwave thermal imaging camera IR Flir SC 5000 (image frequency 608 Hz with a 0.02-K sensitive cooled mediumwave InSb detector).

The high strain rate compression tests were performed on a single-stage powder gun. The powder gun can accelerate a projectile through the combustion gas of the powder detonation and has already been successfully used to determine the high strain rate behavior of aluminum foam [8] and auxetic structures [25]. The device consists of a breach with sufficient strength, a barrel, and a target chamber. The inner surface of the cylinder was machined with high accuracy by a honing process with an inner diameter of 40 mm. A window for optical observation, terminals for electrical measurements, and a shock absorber for the projectile were installed in the target chamber. The projectile sabot was made of ultra-high-molecular-weight polyethylene (UHMWPE), with a brass weight mounted on the front of it as a projectile driver. The cellular structure was positioned and glued to the brass weight. The influence of air resistance during the tests was negligible since the target chamber was decompressed to a near-vacuum with the vacuum pump. The single-stage powder gun can accelerate a projectile to a velocity of up to 1.5 km/s. However, due to the brittle base material of the auxetic cellular structure, the impact velocity was reduced to 200 m/s by using the appropriately weighted sabot (brass weight), thus achieving the engineering strain rate of more than 10,000 s^{-1} for all analyzed cellular material samples.

The high-speed video camera HPV-1 was used to measure the impact velocity and the deformation behavior of the auxetic cellular structures during the impact (manufactured by SHIMADZU, image acquisition number: 100, maximum resolution: 1 μs). The impact pressure of the auxetic cellular structure on the rigid wall was measured with the PVDF measuring device (Piezo-Film Voltmeter, PVF2 11-, 125EK, Dynasen), as already used in the earlier experiments by Tanaka et al. [7].

In all experiments, an appropriate number of samples was used to obtain a statistically representative average response, never less than three samples for each load case.

6.3 UniPore cellular structures

A novel porous metal with unidirectional pores (UniPore) and constant cross section along the length was developed to further reduce the scattering of mechanical and thermal properties by improving the homogeneity and uniformity of the pore morphology and topology while avoiding softening of the material during production [27,30,31]. This new type of porous structure is produced by first assembling some smaller and thinner copper tubes into a single, slightly thicker outer tube, which is then subjected to explosive compaction [32]. With the help of the explosion compaction technique, it is possible to produce longer samples in the order of several meters.

6.3.1 Geometry and fabrication of specimens

The UniPore structure with longitudinal pores initially consists of an outer tube and thin-walled inner tubes, all made of JIS-C1220 copper, which are welded together with an explosion welding technique. Prior to manufacture, the outer tube with an outer diameter of 30 mm is tightly packed with inner tubes (diameter 3 mm) of a much smaller diameter over the entire length of the sample. The structure of the porous UniPore metal produced by this method has a uniform porous cross section in the longitudinal direction, with the pore length being limited only by the size of the samples.

The inner tubes of different thicknesses (0.2, 0.3, and 0.5 mm) with constant outer tube thickness (1.5 mm) were used to produce samples with three different porosities. The initial porosity was calculated as a theoretical value obtained from the cross section of the samples before compaction. To avoid the complete compression of the inner tube's holes (which will

Fig. 6.1 ***UniPore samples with low (32%,* left*), medium (49%,* middle*), and high (58%,* right*) porosity.*** *(From M. Vesenjak, K. Hokamoto, M. Sakamoto, T. Nishi, L. Krstulović-Opara, and Z. Ren. 2016. Mechanical and microstructural analysis of unidirectional porous (UniPore) copper.* Materials and Design *90: 867–880).*

later form the pores of the structure), the tubes were filled entirely with paraffin. Samples with different porosities (high: 58%, medium: 49%, and low: 32%) were fabricated and are shown in Fig. 6.1. The samples were cut in transverse direction for microstructural and mechanical characterization after the paraffin filler was removed by melting. The deformed boundary ends of the tube were cut off (~5–10 mm at each end). The samples have longitudinally oriented pores, which are separated from each other. The cylindrical compression of the samples also led to a cyclic symmetry and concentric arrangement of the pores (see Fig. 6.1), which can be observed in all samples, regardless of their porosity. The hexagonal pores predominate in the middle of the samples. In contrast, the pore shapes change slightly to a quadrangular and pentagonal shape closer to the outer surface (periphery) of the samples.

The moderate welding condition necessary for the successful formation of the interface is not fulfilled in some regions between the inner pipes due to the different collision inclination angle of the pipes because of their round outer interface. Furthermore, the inner metal tubes with a small diameter and a thin wall are quite difficult to obtain at a reasonable cost.

A new production method for UniPore structures was thus proposed [10]. It is based on rolling of cheap thin metal foil with acrylic spacer bars positioned on a surface and subsequent explosive compaction. As soon as the thin film between the spacers is uniformly accelerated, the welding condition associated with the collision angle is considered stable, similar to conventional explosive welding. The manufacturing process of newly developed unidirectional metallic copper is similar to the one used for original UniPore cellular structures, which involves the preparation of the green specimen and the explosive compaction that triggers the mechanism of explosion welding. Two different thin copper foils (JIS-C1100) with thicknesses of 0.3 and 0.4 mm were used to investigate their suitability for explosion welding and the quality of the welded interfaces for this investigation. Thin foils of other metallic materials can also be applied. The final porosity can be easily adjusted by varying the film thickness, which is approximately equal to the inner wall thickness (t) in the final product, and the size of the acrylic spacers. In comparison to the previously investigated UniPore structures, where paraffin was used in the production process, the new approach with the thin acrylic rods makes it possible that no air bubbles remain in the pipes when filling the inner pipes with liquefied paraffin. In addition, thin copper foil is much cheaper than thin copper tubes. Later in the chapter, Fig. 6.4 shows the porous copper samples with an inner wall thickness of 0.3 and 0.4 mm and an achieved porosity of 0.42 and 0.38, respectively. The specimen with a copper center core and 0.4-mm inner wall thickness were also fabricated with an achieved porosity of 0.31 (also later shown in Fig. 6.4). The samples were cross-cut for subsequent metallographic analysis and compression testing. The average sample diameter was 23.4 mm (standard deviation: 0.4 mm). The height of the compression specimens was 10 mm for transverse loading and 20 mm for longitudinal loading. It should be noted that the external dimensions, the size of the pores, the thickness of the internal walls, and the porosity can be easily adapted to individual application requirements.

6.3.2 Quasistatic loading

The compression behavior of UniPore specimens of different porosities under transverse loading direction concerning pore orientation is shown in Fig. 6.2. The local buckling of the outer tube (at the upper and lower part) appears in the high-porosity sample and is not present in the medium- and low-porosity samples. The buckling deformation mode is significantly reduced by reducing the porosity of UniPore specimens due to local increase in bending and buckling stiffness, resulting in higher stresses in the pipes and higher forces in the bond interfaces between the pipes. Tensile stresses occur in the cell walls in a direction perpendicular to the compression load. In this way, the inner pipe thread prevents the outer wall from bending freely and subsequently from buckling. These transverse tensile stresses consequently lead to the rupture of the adhesive bond between the inner pipes, which is predominant at lower porosities. The first plane

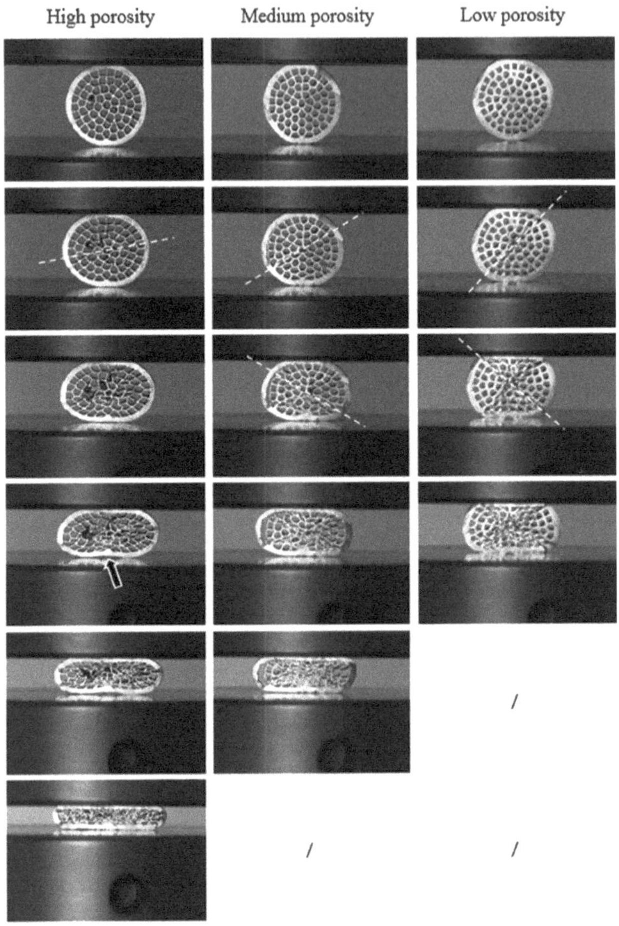

Fig. 6.2 ***Transversal compressive testing of UniPore samples of different porosities (strain increment*** $\Delta\varepsilon \approx 0.15$***).*** *From M. Vesenjak, K. Hokamoto, M. Sakamoto, T. Nishi, L. Krstulović-Opara, and Z. Ren. 2016. Mechanical and microstructural analysis of unidirectional porous (UniPore) copper.* Materials and Design *90: 867–880.*

of failure in the UniPore samples with the highest porosity is almost parallel to the load plates with an inclination of only 10 degrees. The angle of the failure plane for the specimens with medium porosity increases to about 25 degrees, whereas a second symmetrical failure plane appears at an angle of about -25 degrees with further loading, and forces the specimen into a symmetrical global failure deformation mode. By further increasing the thickness of the cell walls (sample with low porosity), it is possible to observe an increase in the angle of the first failure plane to 40 to 45 degrees

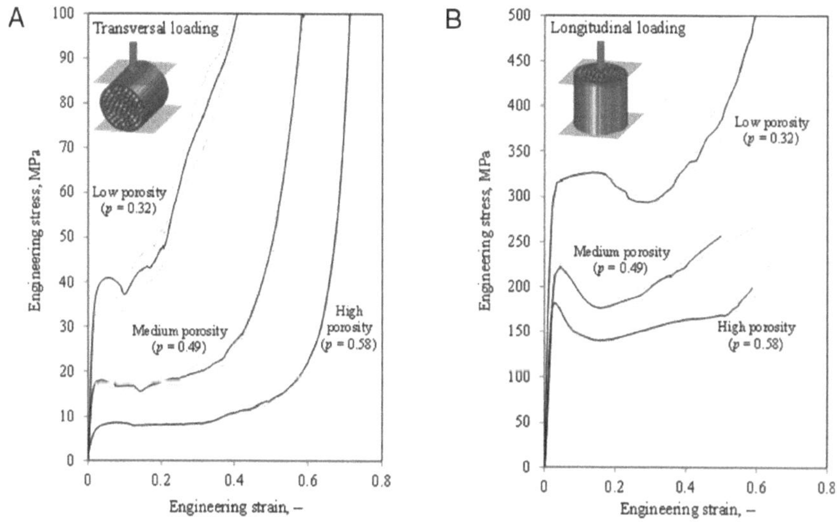

Fig. 6.3 ***Engineering stress–strain relationship of UniPore specimens subjected to transversal (A) and longitudinal (B) compressive loading.*** *(From M. Vesenjak, K. Hokamoto, M. Sakamoto, T. Nishi, L. Krstulović-Opara, and Z. Ren. 2016. Mechanical and microstructural analysis of unidirectional porous (UniPore) copper.* Materials and Design *90: 867–880).*

and the occurrence of a second symmetrical failure plane under further loading. Complete compaction of UniPore medium- and low-porosity samples could not be achieved due to load limit (50 kN) of the testing machine used for the transverse compression test.

The corresponding engineering stress–strain relationships of UniPore specimens for the lateral and longitudinal load directions are shown in Fig. 6.3. The relationships can be divided into four main areas in both loading cases: (1) the initial quasilinear stress increase (quasielastic area); (2) the transition zone in which the cell walls begin to bend, buckle, and deform plastically locally (reaching the initial stress maximum); (3) the stress plateau with almost constant stress; and (4) densification, in which the stress increases dramatically due to contact between the tube walls.

In the case of transverse loading (see Fig. 6.3A), it is clear that porosity significantly influences the global behavior, especially plateau stress (there is an approximately 100% increase between different UniPore samples) and densification strain. Higher stresses at lower porosities can also be observed due to the limited lateral deformation of the outer tube and due to the higher stiffness of the inner structure. It is evident that the densification strain shifts toward the lower strains (from 0.8 for high porosity to 0.19 for low porosity) by reducing the porosity. It is also interesting to note that the smoothness of the plateau stress increases with increasing porosity. Although the plateau stress is constant at high porosity, a stress drop after the initial stress maximum can be observed for samples with medium and low porosity. The stress drop is 17.8% higher for a sample with low porosity than for a sample with medium porosity. It can also be observed that the scatter of results increases due to the reduction of porosity. In the case of the UniPore structure with high porosity, the dispersion of results is almost negligible. As expected, the specimens subjected to longitudinal loading (see Fig. 6.3B) can absorb much higher forces and offer higher mechanical resistance due to the simultaneous pore orientation and the different compression set mode (which confirms the anisotropy of the structure). It can be observed in all samples that a stress drop follows an initial stress peak. A significant difference in the plateau region between the transverse and longitudinal loading must also be taken into account. While the stress drop (unsteady behavior) increases due to the decrease in porosity during transverse loading, the opposite can be observed for longitudinal loading.

The compressive behavior of the rolled UniPore samples of different porosity under longitudinal loading concerning the pore orientation is shown in Fig. 6.4. The number of folds and the position of the first fold depends on the geometric and material properties. The empty tube begins to deform in the weakest region, creating the first fold. After complete compression, the tension stress increases again, which leads to the formation of the next fold. The process is repeated, with the stress peaks in the stress–strain diagram representing the number of folds.

It is expected that the manufactured rolled UniPore structure will exhibit strong orthotropy under mechanical stress due to the unidirectional pore orientation. Therefore, the compression behavior of the sample was analyzed in transverse (sample length: 10 mm) and longitudinal direction (sample length: 20 mm) with respect to the pore direction. The transverse and longitudinal compression deformation behavior of fabricated specimens with three different porosities is shown in Fig. 6.4. In the event of a transverse load, the inner pipe threads prevent the outer wall from bending freely and subsequently buckling. Consequently, these transverse tensile stresses lead to a break in the bond between the cell walls at a deformation of 0.5. During longitudinal loading, an asymmetrical fold was formed approximately in the middle of all specimens during the entire deformation process (which occurred after uniform deformation at a deformation of approximately 0.15).

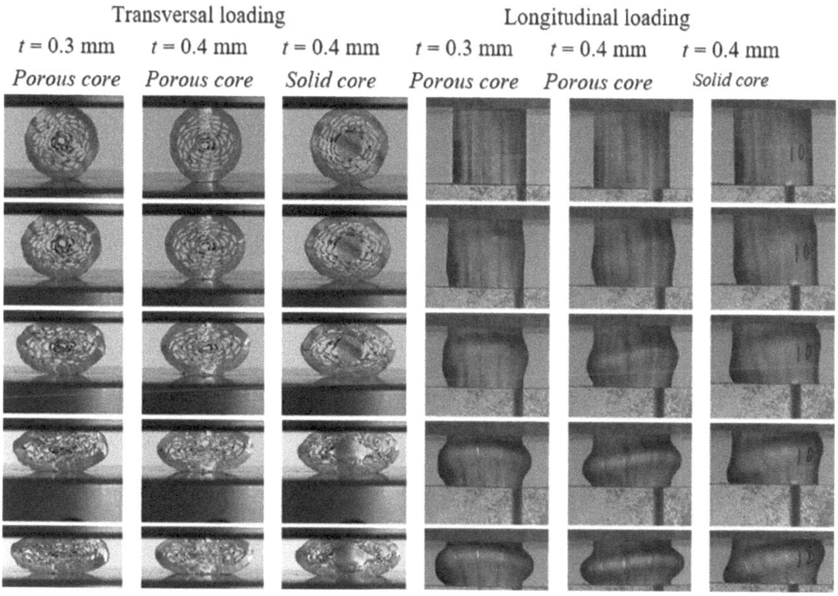

Fig. 6.4 ***Transversal and longitudinal compressive behavior of rolled UniPore structures (strain increment $\Delta\varepsilon \approx 0.15$).*** *(From M. Vesenjak, K. Hokamoto, S. Matsumoto, Y. Marumo, and Z. Ren. 2016. Uni-directional porous metal fabricated by rolling of copper sheet and explosive compaction.* Materials Letters *170: 39–43).*

Only one nonsymmetrical fold was formed approximately in the middle of all UniPore samples (high, medium, and low porosity) by the entire deformation process (see Fig. 6.4), which occurred after uniform deformation at an elongation of about 0.1. This again corresponds to the diagrams, where a single fold is expected after the thickness/diameter ratio exceeds 0.1 (taking length/diameter = 1 into account). The inclination of the folding plane increases from 10 degrees for the samples with high porosity to 35 degrees for the samples with low porosity. The fold size increases by reducing the porosity due to the internal structure and the thickness of the inner cell walls.

The engineering stress–strain response of rolled UniPore specimens is shown in Fig. 6.5, where it is also compared with the original UniPore structures with similar porosity [27]. The figures show a strong anisotropy and the typical compression behavior of porous materials in a transverse direction. The change in film thickness affects the stiffness of the structure, but it does not change the shape of the response. It can be seen that the samples with thicker film experience higher stiffness and lower densification strain.

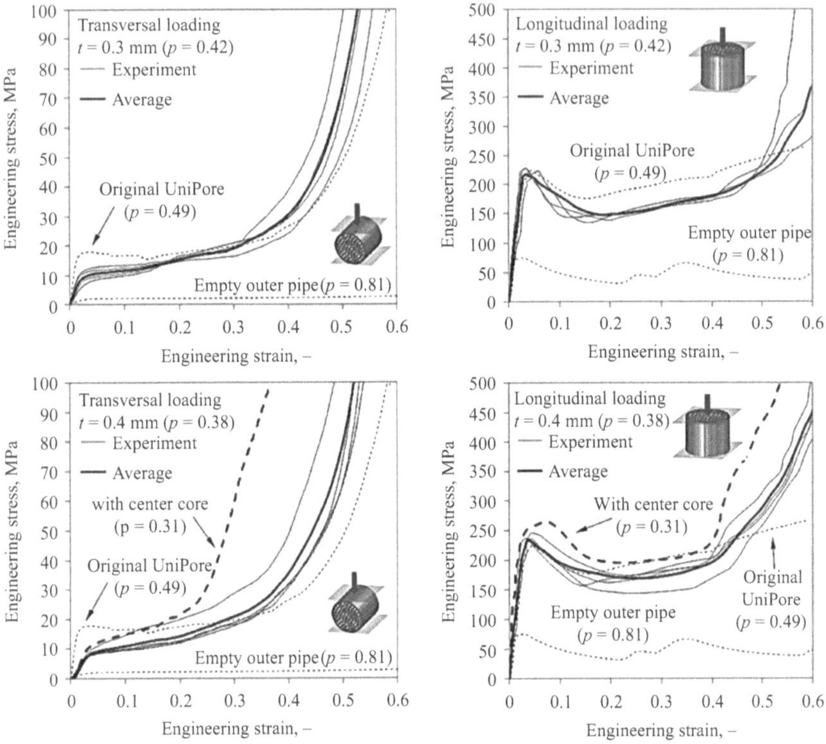

Fig. 6.5 ***Engineering stress–strain relationships for transversal and longitudinal compressive loading directions.*** *(From M. Vesenjak, K. Hokamoto, S. Matsumoto, Y. Marumo, and Z. Ren. 2016. Uni-directional porous metal fabricated by rolling of copper sheet and explosive compaction.* Materials Letters *170: 39–43).*

6.3.3 High strain rate loading

UniPore samples for high strain rate testing were cut from the fabricated UniPore pipes to obtain cubical samples with dimensions of 17.3 mm × 17.3 mm × 17.3 mm. The deformation behavior of UniPore samples under high strain rate loading is shown in Fig. 6.6. The cubic copper UniPore sample hit the rigid wall perpendicular to the direction of the pores. Fig. 6.6 shows the high-speed camera images of the sample, which provide information about the deformation behavior during the impact. The specimen was fully compressed during the impact. The deformation started from the target side and spread through the specimen. The starting time (0 μs) was defined as the time at which the sample first came into contact with the rigid wall (target). The first impact flash was observed at 4 μs, followed by

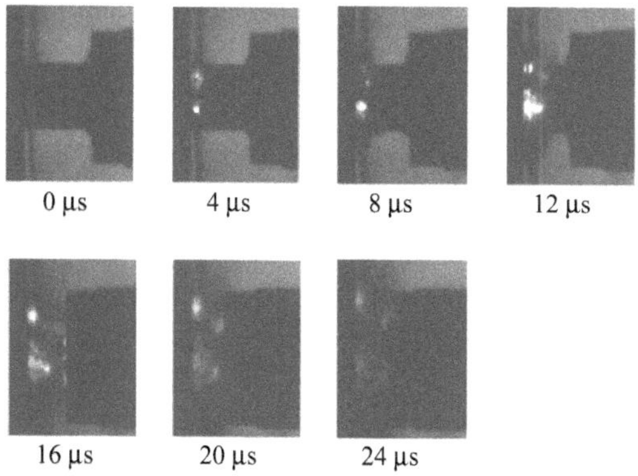

Fig. 6.6 *Deformation process of UniPore samples observed through the high-speed camera.* *(From M. Nishi, S. Tanaka, M. Vesenjak, Z. Ren, and K. Hokamoto. 2020. Experimental and computational analysis of the uni-directional porous (UniPore) copper mechanical response at high-velocity impact.* International Journal of Impact Engineering *136: 103409).*

continuous flashes during the impact. The flashes are related to the formation of the metal jet caused by closing pores and propagating outward in relation to the pores.

Pressure curves derived from the PVDF pressure sensor measurement and the results of the computational simulation are shown later in Fig. 6.9. Both results show similar behavior with two pressure peaks. The first peak occurred when the sample first hit the target, whereas the second pressure peak was observed when the small pores on the surface of the sample were fully compressed.

6.3.4 Computational modeling

ANSYS AUTODYN software was used for computational simulation to validate the manufacturing process of the UniPore cellular structures. In the experimental setup presented in the work of Hokamoto et al. [26], computational simulations were carried out for samples with low, medium, and high porosity, which were achieved by changing the thicknesses of the inner tubes from 0.5, 0.3, and 0.2 mm, respectively. In this study, the copper tubes, which consist of an outer tube (30-mm diameter, 1.5-mm wall thickness), 63 inner tubes (3-mm diameter), and paraffin embedded in the inner tubes, were modeled using the Lagrangian finite element model. The multicomponent Eulerian solver was used to model the explosive PAVEX. The empty space around the explosive was also considered with the same numerical grid. The interactive Euler–Lagrange boundary condition implemented in ANSYS AUTODYN was applied to the outer surfaces of the outer pipe.

The computational simulations were performed using a planar 2D model with a cross section of the experimentally manufactured UniPore assembly [21]. The Mie–Grüneisen equation of state (EOS), which is based on the Hugoniot shock law and the Johnson-Cook constitutional law, was applied. In addition, the Mie–Grüneisen EOS without the material strength model was used to model the paraffin behavior. The results are in excellent agreement with the experimental results [21], which justifies the possibility of using simulations to predict the final shape of porous samples of different dimensions and porosities.

The high strain rate loading of UniPore structures was analyzed with computational simulations performed using the Lagrange–Lagrange interactions in the ANSYS AUTODYN code [28]. In the model, all materials (i.e., steel, copper, and UHMWPE) are described with the Lagrangian mesh. The projectile was fired at an initial velocity in the direction perpendicular to the target, and a fixed boundary condition was applied to the anvil. The anvil and target were joined, and frictionless contact between the target and sample was prescribed. Three-dimensional hexahedral elements were used to mesh all parts. An unstructured grid with 175,938 nodes and 149,930 elements was used for the UniPore sample. The image data of the front view of the sample was imported into CAD software (Solidworks) to model the sample, whereas the pores were represented with circles with a radius of 1.5 to 1.6 mm at the same position as the hole positions of the image data. Structured grids with 21,168, 7056, 7056, and 10,648 nodes were applied to the anvil, target, plate, and sabot, respectively.

The material of the anvil and the target is 4340 steel, whereas copper was used for the sample and plate. The Mie–Grüneisen and Shock–Hugoniot EOS and the Johnson–Cook constitutive law, which allows the consideration of work-hardening and strain rate effects, were applied to 4340 steel and copper. A Mie–Grüneisen-type EOS of the Hugoniot equation of the shock was used for UHMWPE.

The velocity of the projectile was calculated by image analysis during the free flight of the projectile at 387.5 m/s. Consequently, the impact velocity was set as 387.5 m/s in the computational simulation. The simulated deformation process is shown in Fig. 6.7. Similar to the experimental results, the cubic copper sample began to deform at the

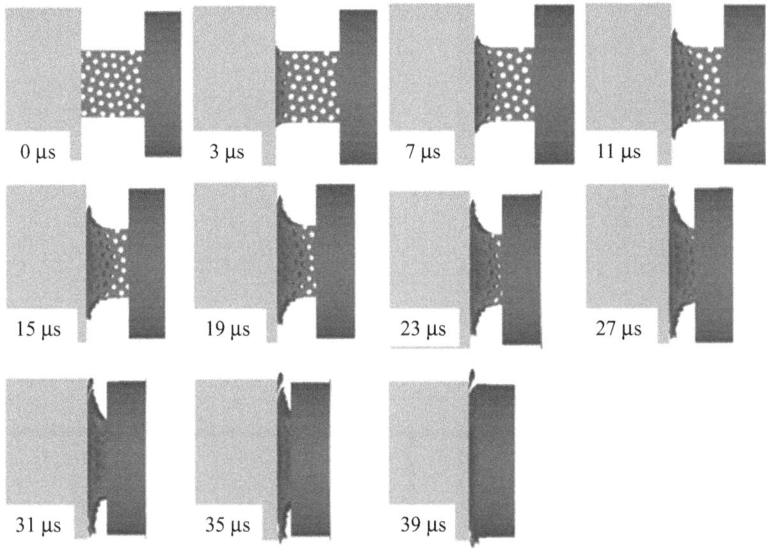

Fig. 6.7 ***Deformation process predicted through computational simulation. The sabot is not displayed.*** *(From M. Nishi, S. Tanaka, M. Vesenjak, Z. Ren, and K. Hokamoto. 2020. Experimental and computational analysis of the uni-directional porous (UniPore) copper mechanical response at high-velocity impact.* International Journal of Impact Engineering *136: 103409).*

target site and was fully deformed at 39 μs after the first contact of the sample with an anvil.

Fig. 6.8 shows the final experimental and computationally predicted deformation state of the cubic UniPore specimen after the dynamic compression test. The specimen expanded in all directions (perpendicular to the direction of loading) and shows a deformed circular structure, with a good agreement between experimental and computational results.

Although the deformed sample volume predicted by the simulations is larger than the one of the test result, the upper and lower ends of the test sample were further analyzed. The cross section of the experimentally tested sample shown in Fig. 6.8A was etched to visualize the microstructure, where the closed pore parts are particularly visible. As observed, the pores are closed and considerably stretched.

Pressure histories derived from the PVDF gauge measurement and computational simulation are shown in Fig. 6.9. Both results show similar behavior with two pressure peaks. The first peak occurred when the sample first hit the target, whereas the second pressure peak was observed when the small pores on the surface of the sample were fully compressed.

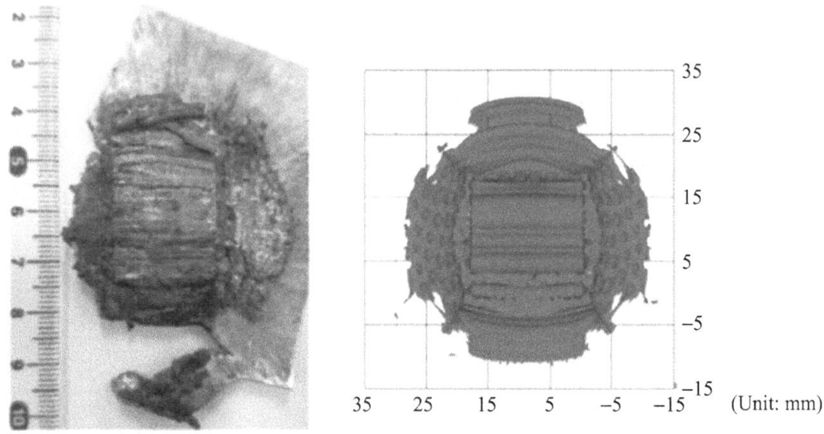

A B

Fig. 6.8 ***Experimental (A) and computational (B) result of the final deformation state (right-side view). The sabot and plate are not displayed.*** *(From M. Nishi, S. Tanaka, M. Vesenjak, Z. Ren, and K. Hokamoto. 2020. Experimental and computational analysis of the uni-directional porous (UniPore) copper mechanical response at high-velocity impact.* International Journal of Impact Engineering *136: 103409).*

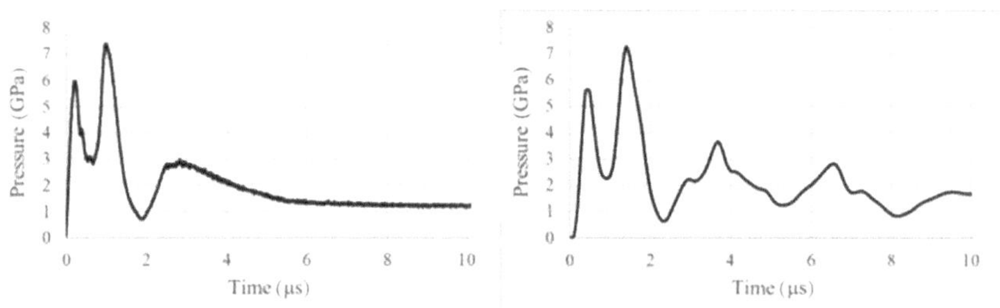

A B

Fig. 6.9 ***Experimental (A) and simulation (B) results of pressure history on the target.*** *(From M. Nishi, S. Tanaka, M. Vesenjak, Z. Ren, and K. Hokamoto. 2020. Experimental and computational analysis of the uni-directional porous (UniPore) copper mechanical response at high-velocity impact.* International Journal of Impact Engineering *136: 103409).*

Although the first pressure peak in the simulation is slightly lower, the second pressure peak coincides with the experimental and computational results. These results also agree well with the prediction using the impedance matching technique [28].

6.4 Open- and closed-cell foams

Most of the foams can be divided in two major groups: open- and closed-cell foams. Cellular structures (foams) of each of those groups are presented in this chapter regarding their fabrication, experimental testing, and computational simulations.

The mechanical properties of metallic open-cell materials (Fig. 6.10) primarily depend on the properties of the base material, morphology, and topology of the cells. The influence of the base material on the properties of metal foams is well studied, and there are many empirically established relationships linking the properties of the base material and morphology with the properties of the cellular material. Metals and alloys used for metal foams must have a low density to maintain the advantage of low relative density over conventional solid materials. For this reason, the most commonly used metals for cellular materials are aluminum, magnesium, titanium, and their alloys [29].

The influence of cell morphology on the mechanical properties of regular and irregular open-cell materials has been investigated by many authors using the representative unit cell. It was assumed that the behavior of the cell material could be described well enough with a single, geometrically correct cell. Since the cell geometry of real open-cell foams differs from the geometric regularity, many authors have also investigated the cell morphology of real foams [19]. However, it should be taken into account that solid foams differ significantly from liquid foams because their base materials have entirely different physical properties.

The mechanical behavior of closed-cell aluminum foams has been thoroughly characterized under free and laterally constrained compression loading conditions [30]. The same foams were also introduced into tubes (*ex-situ* foam-filled tubes) and tested under compressive and flexural stress

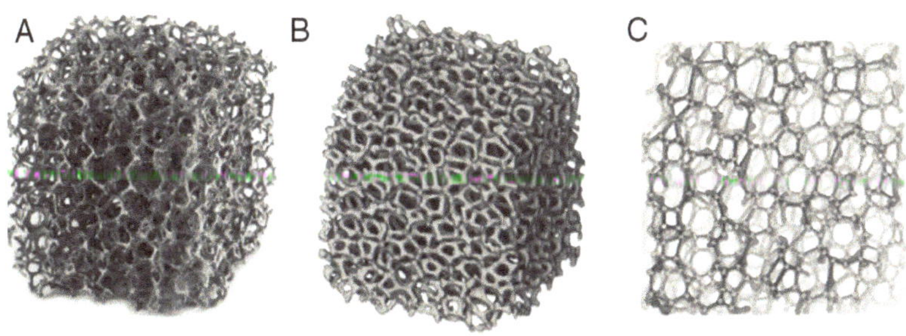

Fig. 6.10 ***Aluminum open-cell foam sample (A) and foam surface models (isometric view (B) and right view from the top (C)).***

[31,32]. Powder metallurgical foam production also allows the production of closed-cell foams inside the tubes (*in-situ* foam-filled tubes) to achieve increased stiffness through better bonding between the foam filler and the outer tube [7]. The behavior of closed-cell foams under various loading conditions has also been studied in detail [33] using micro-computed tomography (μCT) to track changes in the internal structure during the deformation of closed-cell aluminum foam [34].

The influence of the loading velocity (up to 150 m/s) on the mechanical behavior of additively manufactured lattice materials [25] and closed-cell aluminum foams [35] was analyzed with the split Hopkinson pressure bar apparatus. The pressure behavior of foams made of aluminum alloys at high strain rates has also been investigated and discussed in the work of Deshpande and Fleck [36] and Zheng et al. [37]. Liu et al. [38] investigated the effect of porosity and differences in the deformation pattern of closed-cell aluminum foam under quasistatic and dynamic (split Hopkinson pressure bar) loading.

The change of the deformation mode and the hardening with high strain rate was also observed in computational simulations of metallic foams. Mesoscopic simulations were performed using the μCT scan–based geometric models [38,39]. The arbitrary Lagrange–Eulerian method was also applied to take into account the entrapped air inside the closed-cell foam [40]. A cell-based approach also confirmed deformation localization at higher strain rates [41]. Computational investigations of closed-cell aluminum foam and cell-based dynamic Voronoi lattice behavior showed relationships between cell geometry and loading rate [42,43]. The behavior of closed-cell foams can be successfully described by homogenized computational models, such as crushable foam models, which have been successfully applied to the simulated polymer, auxetic, and aluminum foam cores of composite panels under various loading conditions [44–46]. In addition to closed-cell foams, similar work has been carried out in the field of open-cell foams with regard to multiaxial loading [47] and impact response [48,49].

All of the preceding studies observe that hardening occurs at a high strain rate above a specific (limit) load velocity, which is usually a consequence of inertial effects. A detailed overview of the dynamic compression behavior of cellular materials is given in the work of Sun and Li [50]. As it has already been demonstrated, the deformation mode of a given cellular material or structure changes with increasing load velocity (strain rate). The deformation behavior of cellular materials at different load velocities can generally be divided into three deformation modes: homogeneous, transition, and impact deformation modes [37, 51–53]. These three deformation modes are separated by two critical load velocities.

6.4.1 Geometry and fabrication of specimens

The open-cell foam samples were produced by m.pore GmbH using the investment casting method. The starting point is a porous polymer precursor. The pores are filled with a refractory slurry, which is dried and burnt. During combustion, the precursor pyrolyzes, whereas the slurry hardens and forms the mold for the subsequent investment casting of the metallic matrix. The final step is the removal of the molding material, which results in a porous metallic geometry very similar to the polymer precursor. Its base material is Al99.7%, and the porosity of the structure is 93.9%. Their average pore size is 20 pores per inch. The sample was analyzed using computed tomography with X-rays (Toshiba-Aquilion 64), whereby the 3D model was obtained by computational reconstruction from the set of 2D images (slices) shown in Fig. 6.10.

The closed-cell aluminum foam was produced using the powder metallurgical method [54], which consisted of placing extruded, foamable precursor in a cylindrical stainless steel mold in a preheated oven (MJAmaral, Vale Cambra, Portugal) at 700 °C for 12 min. The cavity of the cylindrical form with an inner diameter of approximately 25 mm and a length of approximately 150 mm was filled by foaming the precursor material with a mass of 61 g of aluminum, silicon (7 wt.%), and titanium hydride (0.5 wt.%). The precursor material was produced by cold isostatic pressing (Schunk Sintermetalltechnik GmbH, Giessen, Germany) of the powder mixture, followed by a horizontal direct extrusion machine (Honsel MN AG, Meschede, Germany), resulting in a rectangular bar with a cross section of 20 mm × 5 mm [9]. The mold with molded foam was removed from the oven and cooled to room temperature. The cylindrical aluminum foam was then removed from the mold made of carbon steel and cut lengthwise to prepare the cylindrical specimens (see Fig. 6.1D) for mechanical testing. The average pore diameter of the resulting closed-cell foams is about 2.65 mm (standard deviation: ~0.87 mm). The average length of the specimens was 23 mm, with average weight of 7.9 g, which results in 75% porosity.

6.4.2 Quasistatic loading

Four cubic samples of the open-cell aluminum foam (40 mm × 40 mm × 40 mm) were subjected to a quasistatic uniaxial compressive load at a

Fig. 6.11 ***Deformation behavior of the open-cell aluminum foam.*** *(From M. Vesenjak, C. Veyhl, and T. Fiedler. 2012. Analysis of anisotropy and strain rate sensitivity of open-cell metal foam.* Materials Science and Engineering*: A 541: 105–109).*

velocity of 0.2 mm/ on the servo-hydraulic testing machine INSTRON 8801 to determine the quasistatic mechanical properties. A typical aluminum foam sample behavior is shown in Fig. 6.11.

Later, Fig. 6.18 shows the engineering stress plotted versus engineering strain of the open-cell aluminum foam under compressive quasistatic loading and its comparison to the computational results. The shape of the curves represents the characteristic compressive behavior of porous materials. After a quasilinear elastic response, the aluminum foam exhibits first buckling, plastic deformation, and collapse of intercellular walls reaching the stress plateau, where the mechanism of buckling and collapse becomes even more pronounced, resulting in large strains at almost constant stress. At high strains, the foam stiffness increases due to densification and converges toward the stiffness of the base material.

The deformation behavior of closed-cell foam is shown in Fig. 6.12. The results of quasistatic compression tests in the form of an observed engineering stress–strain relationship showed a characteristic compression behavior of cellular materials (Fig. 6.13). After an initial quasilinear stress increase, the cell walls begin to bend and buckle, leading to the stress plateau typical for cellular materials [55]. The cell walls begin to break as they

Fig. 6.12 ***Deformation sequence of closed-cell aluminum foam under quasistatic loading conditions (strain increment: 12%).*** *(From N. Novak, M. Vesenjak, I. Duarte, S. Tanaka, K. Hokamoto, L. Krstulović-Opara, B. Guo, P. Chen, and Z. Ren. 2019. Compressive behaviour of closed-cell aluminium foam at different strain rates.* Materials (Basel) *12(24): 4108).*

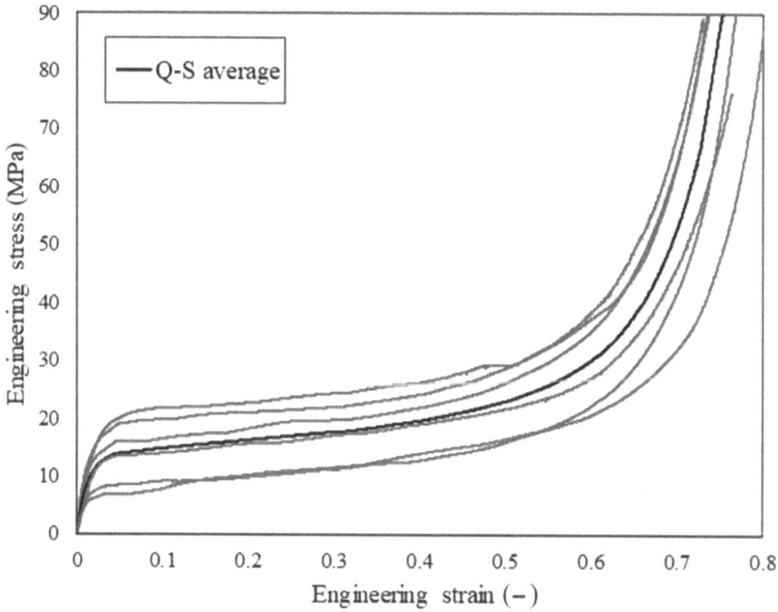

Fig. 6.13 ***Engineering stress–strain relationship of closed-cell aluminum foam under quasistatic (Q-S) loading conditions.*** *From N. Novak, M. Vesenjak, I. Duarte, S. Tanaka, K. Hokamoto, L. Krstulović-Opara, B. Guo, P. Chen, and Z. Ren. 2019. Compressive behaviour of closed-cell aluminium foam at different strain rates.* Materials (Basel) *12(24): 4108.*

continue to deform, and the cells gradually collapse, eventually leading to the densification of the cellular structure. The deviation of the results is a consequence of the variation of the aluminum foam density through the samples (628–713 kg/m^3), which strongly influences the mechanical behavior. Studies have shown that these foams develop imperfections and structural defects (e.g., micropores) during their fabrication, resulting in weaker areas, where the foam begins to deform and develops one or more visible deformation bands perpendicular to the direction of loading [24,56]. The stiffness of the samples is increased by increasing the foam density. Furthermore, the plateau region of these specimens is slightly inclined.

6.4.3 Medium strain rate loading

Fig. 6.14 shows the deformation and failure mechanisms of closed-cell aluminum integral skin foam under quasistatic (see Fig. 6.14A) and dynamic (see Fig. 6.14B) loading conditions. The results show that the specimens start to deform even at lower strains, with the local deformation being

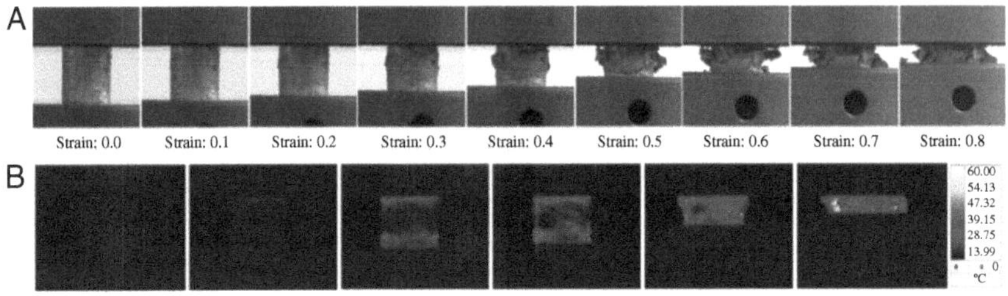

Fig. 6.14 ***Quasistatic (A) and dynamic (B) deformation behavior of closed-cell aluminum alloy foam.*** *(From I. Duarte, M. Vesenjak, and L. Krstulović-Opara. 2016. Compressive behaviour of unconstrained and constrained integral-skin closed-cell aluminium foam.* Composite Structures *154: 231–238).*

distributed at the foam surface, leading to the formation of many deformation bands. The deformation of these materials under axial compressive loads is uneven within the specimens due to the used fabrication process of the closed-cell aluminum alloy foams. Due to the presence of these defects and structural imperfections, such heterogeneous foams usually carry much lower stresses than foams with a regular and homogeneous structure. In addition, the foams investigated in this work were prepared by using several pieces of foamable precursors, which led to the formation of joint areas with an additional increase in defects and structural imperfections within the foam structure. Basically, a foam sample begins to deform in one or more of the weakest regions, where such defects and structural imperfections exist. It is assumed that the foams investigated in this work begin to deform in the joint areas that are formed during the foam formation process. X-ray tomography showed that the formed connection areas consist of several imperfections and structural defects, which lead to stress concentration at which the cracks develop. The deformation is localized in the weak cells, which collapse prematurely due to their lower load-bearing capacity and contribute to an increase in compressive stress, resulting in the formation of a localized deformation band. The foam sample could start to deform at several weakest points and show one or more deformation bands, as was shown in the case of quasistatic loading (see Fig. 6.12). The first deformation bands are observed on the top and bottom of the specimens perpendicular to the direction of loading. The progressive collapse of the cells occurs in localized deformation bands, whereas the cells outside of these deformation bands remain in their original shape. These deformation bands deform until local hardening occurs, causing other deformation bands to collapse entirely. The various deformation bands are distributed within the foam, as shown in

Fig. 6.14. From the IR images (see Fig. 6.14B), it can be seen that the deformation bands are oriented in different directions and inclined at different angles. The deformation of foam is achieved by the formation, propagation, and spreading of strain localization bands throughout the sample. Thermal images (see Fig. 6.14B) show plastification and fracture initiation near contact zones, which continue to spread in foam layers under the outer integral skin, and show that the most critical parts are transition zones between free-form surfaces (inner sample parts) and constrained zones (outer skins). The same behavior is visible for the quasistatic case (see Fig. 6.14A).

6.4.4 High strain rate loading

A high-velocity impact experiment was carried out in which a closed-cell aluminum foam sample was accelerated to 400 m/s with a single-stage powder gun. The velocity of the sample was observed using the high-speed camera and was considered to be constant during the impact. A general stress plateau effect was observed in the reaction of the foam sample, which is typical for cellular material structures when subjected to quasistatic loading. The wave behavior of the aluminum foam sample caused by the collapse of the foam pore edges was measured with a PVDF measuring device.

The results of high strain rate tests in the form of the engineering stress–strain relationship are shown in Fig. 6.15. Again, the response is typical for cellular structures with three distinct differences due to the high-velocity loading: (1) initial stress peak, (2) stress fluctuations during the stress plateau, and (3) abrupt compression (no smooth transition between the stress plateau and the densification region, as is common in the quasistatic loading regime).

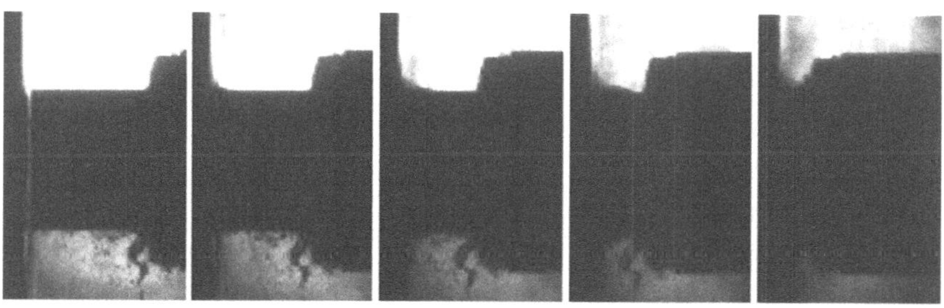

Fig. 6.15 ***Deformation sequence of closed-cell aluminum foam under high strain rate loading (strain increment: 18%).*** *(From N. Novak, M. Vesenjak, I. Duarte, S. Tanaka, K. Hokamoto, L. Krstulović-Opara, B. Guo, P. Chen, and Z. Ren. 2019. Compressive behaviour of closed-cell aluminium foam at different strain rates.* Materials (Basel) *12(24): 4108).*

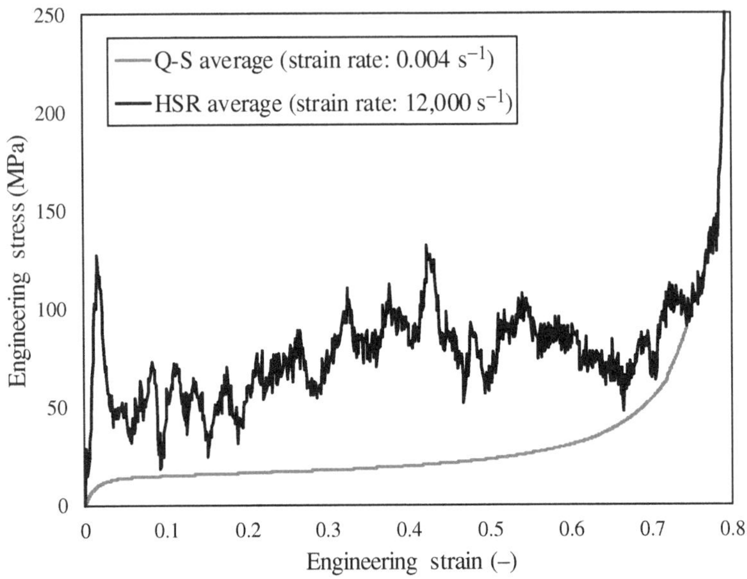

Fig. 6.16 ***Comparison of average quasistatic*** **(Q-S)** ***and high strain rate*** **(HSR)** ***results of closed-cell aluminum foam.*** *(From N. Novak, M. Vesenjak, I. Duarte, S. Tanaka, K. Hokamoto, L. Krstulović-Opara, B. Guo, P. Chen, and Z. Ren. 2019. Compressive behaviour of closed-cell aluminium foam at different strain rates.* Materials (Basel) *12(24): 4108).*

The deformation pattern recorded with the high-speed camera is shown in Fig. 6.15, where it can be observed that the deformation mode changed to impact mode compared to the quasistatic reaction. Consequently, most of the deformation takes place at the impact surface between the rigid wall and the sample until the sample is fully compressed.

The comparison between the quasistatic and high strain rate responses is shown in Fig. 6.16. A significant strain rate hardening effect can be observed in the case of high strain rate loading, which is mostly the consequence of the changed deformation mode and inertia effects.

6.4.5 Computational modeling

The correct geometric representation of fabricated aluminum open-cell foam samples requires a sufficiently high resolution of the μCT scan to capture geometric details and allow for accurate segmentation of the metallic phase. To avoid enormous computing times, the size of the computational model was reduced to 20 mm × 20 mm × 20 mm, which corresponds to one-eighth of the sample size used in the experimental tests. The geometric convergence is ensured during the preliminary analysis of the foam refinement. Based on the results of the network sensitivity analysis, foams with at least 130,000 fully integrated tetrahedral elements were selected for subsequent simulations.

The material data for the aluminum foam base material were defined with the elastic–plastic constitutive model; the strain rate sensitivity was considered with the Cowper–Symonds constitution law.

The boundary conditions combine a displacement-controlled load represented as a rigid wall on the opposite sample surfaces, which simulates a uniaxial compressive load. Three different cases were analyzed separately for each coordinate direction (x, y, and z). This enables anisotropy analysis—that is, the observation of the change in quasistatic and dynamic behavior with respect to the direction of loading. All nodes of the aluminum foam were included in automatic single-surface contact, also taking into account the friction. The computational model was analyzed with the engineering software LS-DYNA MPP 971 R5.0 on the basis of the implicit and explicit integration scheme. For the quasistatic load, an implicit time integration scheme was used, which allows large time steps (solution increment) compared to the dynamic simulation. In the case of dynamic loading, the explicit time integration scheme with sufficiently small time steps was applied. In explicit finite element simulation, large simulation time steps shorten both calculation time and numerical stability.

Fig. 6.17 shows the deformation of the foam with contours representing von Mises stresses during loading in the z-direction. A gradual and uniform

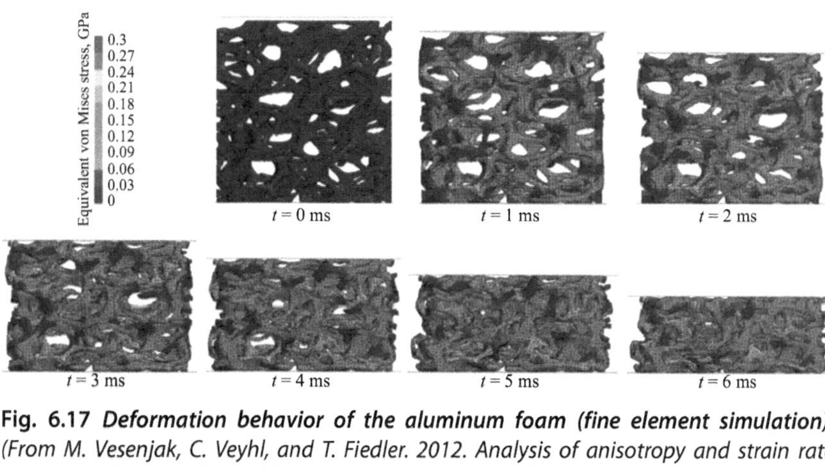

Fig. 6.17 ***Deformation behavior of the aluminum foam (fine element simulation).*** *(From M. Vesenjak, C. Veyhl, and T. Fiedler. 2012. Analysis of anisotropy and strain rate sensitivity of open-cell metal foam.* Materials Science and Engineering: A *541: 105–109).*

deformation of the foam is observed. The results are shown for a strain rate of 100 s^{-1}. However, the deformation behavior does not significantly depend on the strain rates for the values considered up to 1000 s^{-1}. At higher load velocity, a more localized deformation of the foam is expected due to a more substantial inertia influence and shock wave propagation.

The results of the computational simulations were evaluated using the normal forces obtained from the contact area between the metal foam and the rigid wall. The average engineering stress was calculated by dividing the sum of the normal contact forces by the initial cross section. The engineering strain was determined by the compression of the specimen due to the displacement load. The diagrams in Fig. 6.18 show the comparison between the quasistatic experimental measurements and the computational results for a quasistatic load. It should be mentioned here that all experimental tests were performed in the *y*-direction. In Fig. 6.18, the elastic–plastic transition zone is clearly visible. The mechanical properties of the analyzed metallic open-cell foam are anisotropic, and the maximum stresses are observed in the *x*-direction. Nevertheless, the reaction in the *x*- and *y*-direction is very similar. Therefore, the material appears to have been orthotropic, which corresponds to the manufacturing process. From the comparison between the computational and experimental results, an excellent agreement can be observed. The correlation between the experimental and computational results is even more impressive for global behavior up to very large loads (see Fig. 6.18). This confirms the validity of the computational model.

In the case of closed-cell foams, the computational model with the applied homogenized material model (crushable foam) with volumetric

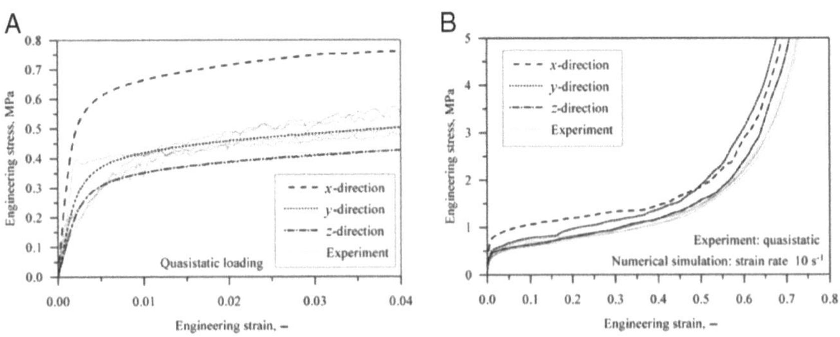

Fig. 6.18 ***Behavior of the aluminum foam under quasistatic (A) and dynamic (B) loading.*** *From M. Vesenjak, C. Veyhl, and T. Fiedler. 2012. Analysis of anisotropy and strain rate sensitivity of open-cell metal foam.* Materials Science and Engineering: A *541: 105–109.*

hardening was built in Abaqus finite element software, where the explicit solver was used for analysis.

Due to the axial symmetry of the specimens, axisymmetric boundary conditions could be applied in the computational model (see Fig. 6.20). The load plate (top) and the support plate (bottom) were modeled with interaction, whereby all edge nodes at the top and bottom edge were connected to two reference points by kinematic coupling. These two reference points were then given boundary conditions. The load plate was specified with a constant velocity of 1 m/s (quasistatic test) and 270 m/s (high strain rate test). The higher calculated load velocity compared to the experimental quasistatic velocity was determined by comparing the reaction forces at the load and support plate, which should be identical if no inertial effects are present. The support plate has a limited displacement in the y-direction and a limited rotation around the z-axis, whereas the loading plate has only limited rotation around the z-axis.

The finite element mesh consists of 299 linear square elements (type CAX4R) with an average global size of 1 mm. The appropriate finite element size was determined during the convergence study with three different finite element sizes.

The crushable foam constitutive model parameters were determined with the optimization algorithm, whereby first the quasistatic experiment and the computational responses were compared (Fig. 6.19). After optimizing the material parameters, the same material model was used for the high strain rate load. Good agreement can be observed for both quasistatic and high strain rate tests. The computational model was thus successfully validated. The agreement between the experimental and computational responses is very good for both analyzed strain rates (see Fig. 6.19), and the discrepancy can only be observed at lower strains of the high strain rate reaction, where only the first stress peak could not be observed in the computational model due to the initial boundary conditions. The stress peak is a consequence of the collision in the experiments and represents a typical reaction of the structures during the initial phase of the impact but does not influence the global behavior afterward [25].

The deformation behavior under quasistatic and high-strain load conditions is shown in Fig. 6.20. Under quasistatic load, a uniform deformation of the sample can be observed. The change in the deformation mode can be observed in the case of high strain rate loading, where the deformation is localized at the impact surface between the load plate and the specimen.

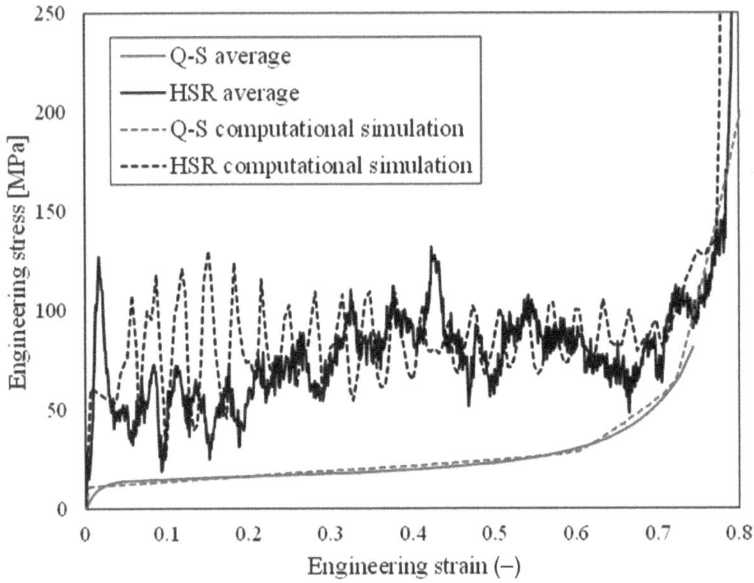

Fig. 6.19 ***Comparison of experimental and computational results of closed-cell aluminum foam behavior at quasistatic (Q-S) and high strain rate (HSR) loading.*** *(From N. Novak, M. Vesenjak, I. Duarte, S. Tanaka, K. Hokamoto, L. Krstulović-Opara, B. Guo, P. Chen, and Z. Ren. 2019. Compressive behaviour of closed-cell aluminium foam at different strain rates.* Materials (Basel) *12(24): 4108).*

Fig. 6.20 ***Equivalent plastic strain evolution in closed-cell aluminum foam during quasistatic (A) and high strain rate (B) loading (strain increment: 15%).*** *(From N. Novak, M. Vesenjak, I. Duarte, S. Tanaka, K. Hokamoto, L. Krstulović-Opara, B. Guo, P. Chen, and Z. Ren. 2019. Compressive behaviour of closed-cell aluminium foam at different strain rates.* Materials (Basel) *12(24): 4108).*

6.5 Auxetic structures

The cellular structure can be designed in a way to exhibit a negative Poisson ratio under mechanical loading—that is, the structure tends to expand in the lateral direction when subjected to longitudinal tensile loading and vice versa in case of compressive loading (see Fig. 6.2). The auxetic cellular structures thus experience significant volume changes under loading, which is very desirable for specific applications, such as crash absorbers, protective padding, stents, and structural elements. Such behavior is made possible by a carefully designed 2D or 3D hinge- or sheet-like cellular skeleton with predefined geometry [57,58]. The auxetic cellular metamaterials offer extraordinary mechanical properties not achievable by any other conventional material—for instance, unique deformation behavior, variable stiffness, density, large mechanical energy absorption through deformation, nonintuitive bending behavior, and increased shear resistance. Their cell shapes change rapidly through the entire cellular structure under loading, thus efficiently spreading the loading through the entire material structure. In case of an impact on a small part of the auxetic cellular metamaterial, the loading energy dissipates through its entire structure. The term *auxetic materials* (from the Greek *auxetos*: it can be increased) was first introduced in 1991 by Evans et al. [15].

The natural auxetic materials are found in α-cristobalite [59], biological tissues (skin) [60], and pyrolytic graphite [61]. Artificially produced material structures with similar behavior were first used in the 1960s in practical application as the moderator core of a Magnox nuclear reactor [62]. The structure has been designed to withstand the horizontal forces that occur during an earthquake while allowing the structure to move freely to accommodate thermal movement between the graphite core and the steel support structures at the edges of the core. The development of this structure was carried out to solve a practical problem, and the auxetic behavior was only a by-product of the design.

The systematic investigation of auxetic materials began in 1985 with the analysis of a 2D-reentrant auxetic structure by Almgren [63], which is based on analytical expressions that were presented for conventional honeycombs [64] a few years earlier. In 1987, the first auxetic material was produced by Lakes and colleagues by converting conventional open-cell foam [65]. Later, auxetic structures were also produced from polymers [21–23]. Further progress has been made with the introduction of composite structures that have a negative Poisson ratio [66]. The next promising method to produce more complex and more extensive auxetic structures is additive manufacturing.

It is not always necessary to have a porous microstructure (as in cellular materials) to induce the auxetic behavior of the material structure [66]. Changes in the microstructure (molecular auxetic materials) or composite structures with rotating units can also exhibit auxetic behavior [67].

Most of the auxetic structures mentioned previously are highly anisotropic and therefore in most cases are not well suited for general structural applications. For this reason, a fascinating development of novel cellular 3D auxetic materials is under way at several research centers, and thus some promising auxetic materials are expected to be introduced.

Most auxetic materials can be divided into three main groups: auxetic honeycombs, auxetic microporous polymers, and auxetic composites. Two different types of auxetic honeycombs will be presented in this chapter.

Experimental studies on the deformation behavior of auxetic cellular materials have so far concentrated mainly on the quasistatic elastic behavior of uniform auxetic structures at very small strains [68] and basic bending testing [69]. The basic uniaxial static compression and tension tests of simple auxetic structures have been used in several other studies [70–75]. There have been some limited experimental studies of the dynamic response of auxetic cellular structures at different strain rates and scales [76–79]. They were mostly concerned with structures of the same geometry, porosity, direction of loading, or strain rate in a particular experimental experiment. As such, they do not provide sufficient data for a proper comparison and evaluation of the parameters influencing their mechanical behavior.

Several computational studies were also conducted to analyze the different dynamic behavior of auxetic cellular structures as parts of composite structures under dynamic loading [80,81], blast loading [82], and ballistic loading [83]. These studies were mainly concerned with the global response of the composite structures to specific loading conditions, whereas the behavior of the nucleus (auxetic cellular structure) itself was not specifically evaluated for the case of different geometric parameters and loading conditions.

The preceding studies mainly present the results of quasistatic and limited dynamic analyses of various auxetic cellular structures produced with additive manufacturing technologies. In this chapter, the influence of load direction and porosity of auxetic cellular structures (two different types: inverted tetrapods and chiral) on their mechanical behavior at different strain rates is experimentally determined. The experimental observations are used to validate the finite element computational models, which provide the means for a more detailed analysis of the underlying deformation mechanism and failure of the analyzed structures. Computational models are validated with the help of IR thermography, which provides a detailed insight into the development of the plastic deformation zones of the auxetic cellular structure under dynamic load.

6.5.1 Geometry and fabrication of specimens

The auxetic cellular structures are built up from inverted tetrapods and chiral auxetic structures. The unit cells of inverted tetrapods are stacked in layers and then into the layered 3D auxetic structure [58] (Fig. 6.21A–C). The auxetic cellular structures analyzed in this study were constructed as CAD models and produced using the selective electron beam melting method from the Ti-6Al-4V alloy powder at the Institute of Materials Science Technology (WTM) of the University of Erlangen–Nuremberg. The strut thickness depends on various predefined parameters of the selective electron beam melting machine and may differ from the original model CAD. The change in electron beam velocity was used in the production of the investigated samples to achieve different strut thicknesses (i.e., different porosities). Two lengths of auxetic structures built from inverted tetrapods were used in this research: the short specimens, as shown in Fig. 6.21, and the long specimens, which were extended in the X2 direction. In total, 36 short samples (15.6 mm × 19.2 mm × 18.7 mm) and 18 long samples (15.6 mm × 40.5 mm × 18.7 mm) with the same unit cell geometry but different strut thickness (i.e., porosity) were fabricated.

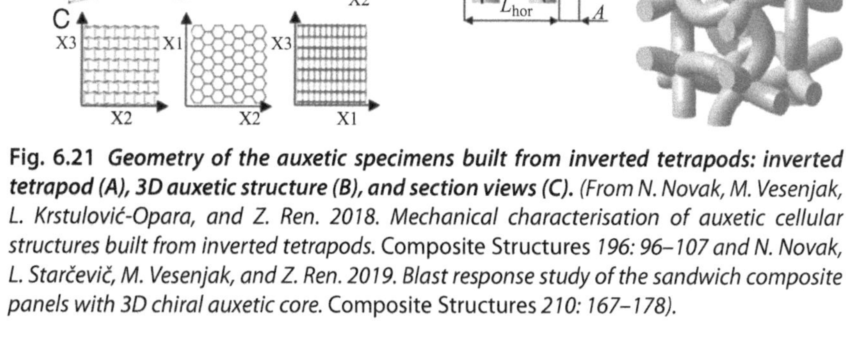

Fig. 6.21 ***Geometry of the auxetic specimens built from inverted tetrapods: inverted tetrapod (A), 3D auxetic structure (B), and section views (C).*** *(From N. Novak, M. Vesenjak, L. Krstulović-Opara, and Z. Ren. 2018. Mechanical characterisation of auxetic cellular structures built from inverted tetrapods.* Composite Structures *196: 96–107 and N. Novak, L. Starčevič, M. Vesenjak, and Z. Ren. 2019. Blast response study of the sandwich composite panels with 3D chiral auxetic core.* Composite Structures *210: 167–178).*

The shape of the chiral auxetic structure unit cell corresponds to the 10th eigenmode of the regular cubic unit cell, which was presented in the work of Körner and Liebold-Ribeiro [84], and studied further by Warmuth et al. [90–92]. The unit cell geometry is defined in Fig. 6.21D and E, where the struts take the shape of the sines, whose turning points meet in the nodal points. The unit cell geometry parameters are the following: nodal distance in vertical (L_{ver}) and horizontal (L_{hor}) directions, amplitude (A), and average strut thickness (d). The specimen dimensions were 55 mm × 25 mm × 25 mm with 5 × 5 × 10 unit cells. The average strut thickness was determined inversely by weight comparison between the fabricated specimen and its CAD model.

6.5.2 Quasistatic loading

The mechanical behavior of auxetic structures has not been systematically characterized up to now due to many influencing parameters that cannot yet be correctly evaluated. This prevents the efficient use of auxetic structures as core components of novel composite structures in various technical applications. There is a clear need for systematic experimental testing of auxetic cellular structures with different geometries and porosities at different strain rates.

The quasistatic compression tests were performed at an engineering strain rate of 0.005 s^{-1}. The observed responses of samples tested in two loading directions (X2 and X3) are significantly different (Fig. 6.22). The

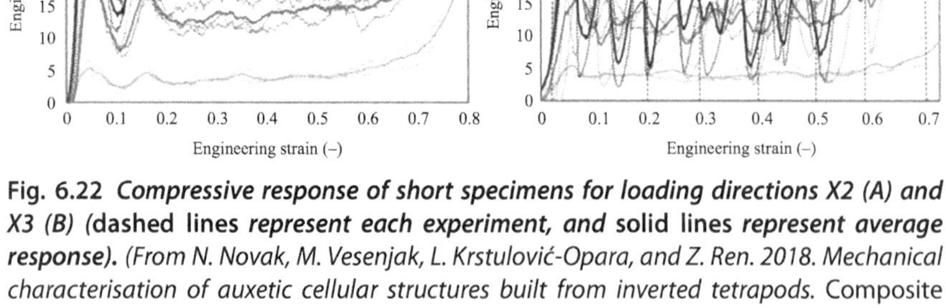

Fig. 6.22 ***Compressive response of short specimens for loading directions X2 (A) and X3 (B)*** **(dashed lines *represent each experiment, and* solid lines *represent average response*).** *(From N. Novak, M. Vesenjak, L. Krstulović-Opara, and Z. Ren. 2018. Mechanical characterisation of auxetic cellular structures built from inverted tetrapods.* Composite Structures *196: 96–107).*

compression behavior of auxetic cellular structures with different porosities, when tested in the X2 direction, is very similar, as can be seen from the shape of the compression response curves in Fig. 6.22A. However, there is a difference in plateau stress between the samples with the two lowest porosities and the sample with the highest porosity. This is attributed to a more significant influence of the manufacturing imperfections on active strut diameter and therefore on the mechanical behavior of specimens with high porosity, which is also observed in loading direction X3 (Fig. 6.23).

In the case of the compressive load in the X3 direction, the observed behavior of samples with three different porosities is very different (see Fig. 6.22B). At lower porosity, the samples show a brittle layer-by-layer collapse behavior, as further shown in Fig. 6.24. However, failure at higher strains has been observed in samples with high porosity, as thin ligaments break at very low strains due to the more significant influence of manufacturing defects (see Fig. 6.23). When we compare samples with medium and low porosity, the response changes with decreasing porosity from moderate to significant layer-by-layer collapse. The average plateau stress of low-porosity samples is approximately similar to that of medium-porosity samples, although the initial yield stress is higher.

As it can be seen in Figs. 6.22 through 6.24, orientation and porosity significantly influence the response of the auxetic cellular structure, which is crucial for its use as a core material in modern composite structures for various applications. Due to manufacturing defects, there is, unfortunately, no linear relationship between the increase in porosity and the increase in plateau stress.

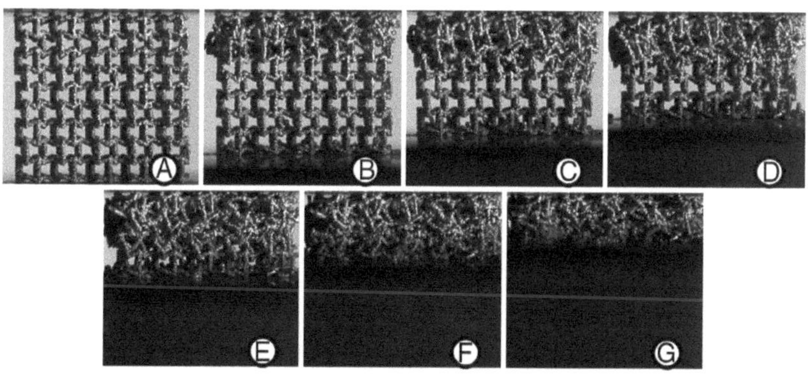

Fig. 6.23 ***Deformation mechanism of a high-porosity specimen (loading direction X3).*** *(From N. Novak, M. Vesenjak, L. Krstulović-Opara, and Z. Ren. 2018. Mechanical characterisation of auxetic cellular structures built from inverted tetrapods.* Composite Structures *196: 96–107).*

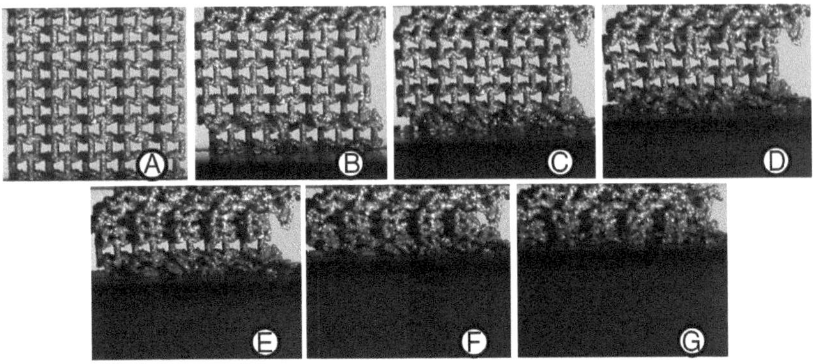

Fig. 6.24 ***Deformation mechanism of low-porosity specimens (loading direction X3).*** *(From N. Novak, M. Vesenjak, L. Krstulović-Opara, and Z. Ren. 2018. Mechanical characterisation of auxetic cellular structures built from inverted tetrapods.* Composite Structures *196: 96–107).*

6.5.3 Medium strain rate loading

The mechanical behavior was further analyzed by means of IR thermography, which provides a more detailed insight into the deformation pattern and particularly the development of the plastic areas in the tested specimens (Fig. 6.25).

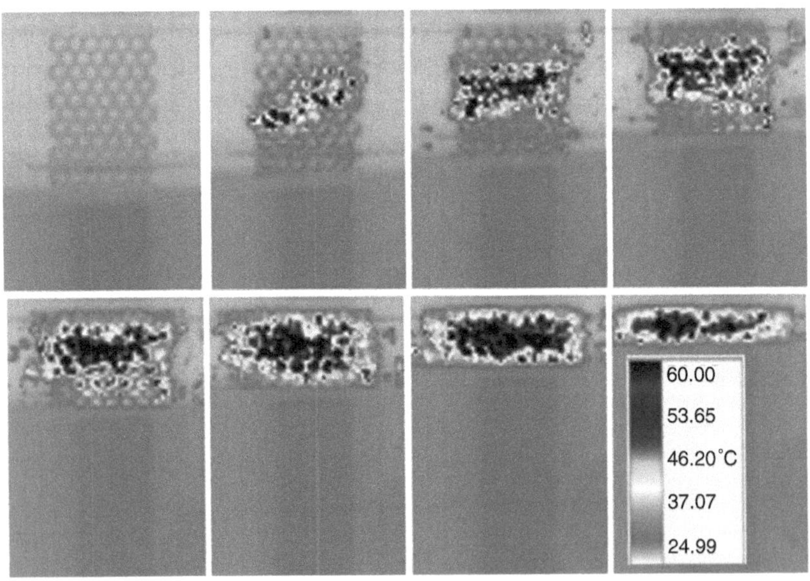

Fig. 6.25 ***Evolution of plastic deformation in short specimens during dynamic loading in the X2 direction (strain increment: 10%).*** *(From N. Novak, M. Vesenjak, L. Krstulović-Opara, and Z. Ren. 2018. Mechanical characterisation of auxetic cellular structures built from inverted tetrapods.* Composite Structures *196: 96–107).*

It can be observed that first one shear band is formed, where the structure starts to collapse. This is different from the case of loading in direction X3 (see Fig. 6.28), where more shear bands are formed.

6.5.4 High strain rate loading

The investigation of the strain rate sensitivity of auxetic cellular structures is directly related to the investigation of conventional cellular materials. In general, the deformation mode of a given material or structure changes with increasing load velocity (strain rate). The deformation behavior can be divided into three deformation modes: homogeneous, transition, and dynamic deformation modes [37,51]. Deshpande and Fleck [36] have shown only a small influence of strain rate on the plateau stress of aluminum alloy foams at high strain rates. However, Tan et al. [52,53] came to the conclusion that there is also a critical impact velocity for aluminum foams, at which the deformation mode changes to "shock mode." Due to these contradictory conclusions, the research reported here focused on the behavior of auxetic cellular structures at high strain rates and the determination of critical velocities by combining experimental tests and computational simulations.

The crushing sequence was analyzed using the digital image correlation of high-speed camera recording. Due to the high strain rates achieved at a loading velocity of 200 m/s (10,000 s^{-1} and 5000 s^{-1}), the impact deformation mode was observed for all analyzed porosities. This means that throughout the tests (up to very large deformations), the main deformation of the specimens was concentrated only in the area, where the specimens hit the solid, rigid plate.

The average pressure (stress) measurements from the gauge at the fixed, rigid measuring plate for long samples of all three porosities are shown in Fig. 6.27. A moving average signal filter (10 data points) was used to smoothen the higher frequency response in the recorded experimental data. A typical compression reaction of the cellular material with almost constant plateau stress and compression at higher strains can be observed in all cases, which is in good agreement with earlier observations [1,4,7,56].

The initial stress peak is a consequence of high strain rate load and pressure data acquisition and is not significant for the overall deformation behavior (see Fig. 6.27). As the difference in porosities is only about 5%, the reactions of long samples are quite similar, with samples with low porosity having a slightly higher stiffness—for instance, plateau stress (80 MPa) compared to samples with medium (70 MPa) and high (50 MPa) porosity. This is also reflected in a slightly lower densification strain of about 0.75 relative to the densification strain for medium- and high-porosity samples of 0.82.

Fig. 6.27 shows the results of the high strain rate tests of long specimens compared to the results of the quasistatic tests (loading velocity: 0.01 mm/s, strain rate: 0.005 s^{-1}) on identical auxetic specimens [17]. The influences of microinertia effects, strain rate sensitivity of the base material, the changed deformation mechanism (from homogeneous [quasistatic] to shock mode) at high strain rates on the initial peak stress, the magnitude of plateau stress, and the densification strain can be clearly observed.

6.5.5 Computational modeling

The simulated crushing behavior of auxetic cellular structures under compressive load in X3 and X2 directions for short and long samples is shown in Fig. 6.28 and later in Fig. 6.30. The deformation patterns in the case of computational models are similar to those observed in experimental tests (see Figs. 6.23, 6.24, and 6.26). Measurement of the longitudinal and transverse displacements showed that the material no longer behaves auxetic at a deformation of more than 0.3% and that the Poisson ratio changes from approximately -0.18 to 0, which is consistent with the results of other researchers [60].

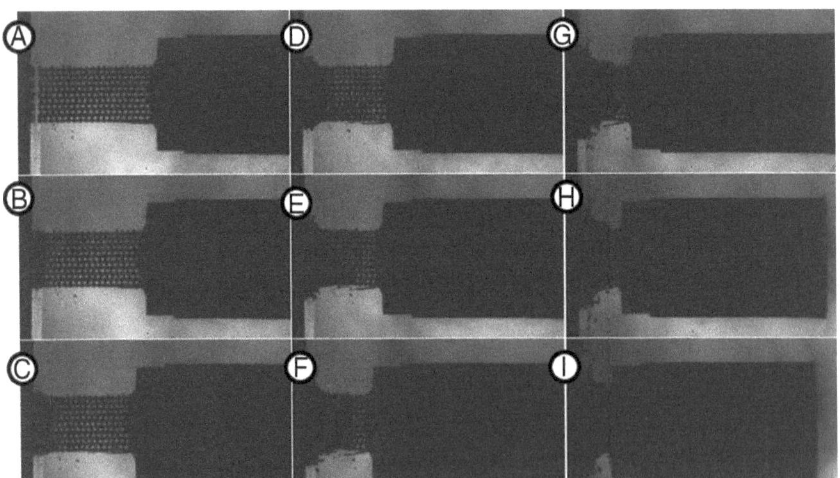

Fig. 6.26 ***Crushing sequence of a long specimen with middle porosity at 200 m/s (strain increment: 10%).*** *(From N. Novak, K. Hokamoto, M. Vesenjak, and Z. Ren. 2018. Mechanical behaviour of auxetic cellular structures built from inverted tetrapods at high strain rates.* International Journal of Impact Engineering *122: 83–90).*

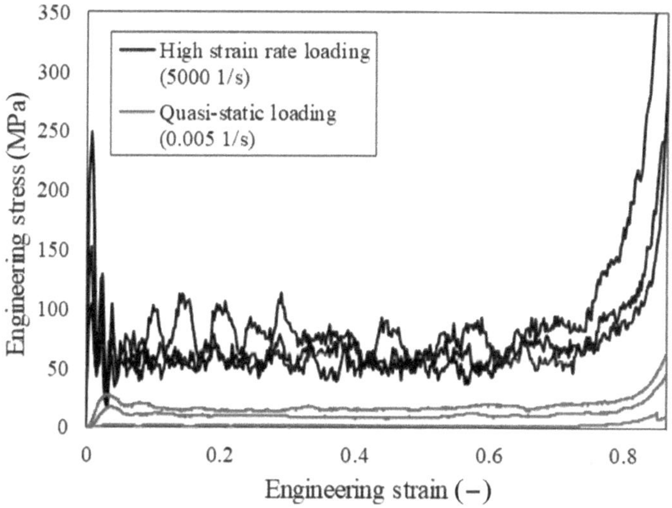

Fig. 6.27 ***Comparison of quasistatic and high strain rate experimental results for long specimens of different porosities.*** *(From N. Novak, K. Hokamoto, M. Vesenjak, and Z. Ren. 2018. Mechanical behaviour of auxetic cellular structures built from inverted tetrapods at high strain rates.* International Journal of Impact Engineering *122: 83–90).*

The computational models were further validated by comparing the deformation behavior of computational models and experimental samples in which the development of plastic deformation was observed by means of IR thermography (see Fig. 6.28).

The reaction force on the bottom plate determined by computational simulations was used to calculate the engineering stress. Fig. 6.29 shows the resulting engineering stress–strain relationship in comparison to the experimental data.

In the case of high strain rate loading, the deformation behavior of short and long samples was first visually compared with the deformation behavior observed in the experiments (see Fig. 6.26). The computational models provide the same deformation pattern as noted in the experiments—that is, a clear impact deformation mode can be observed (Fig. 6.30).

The results from the computational models were further compared with high strain rate experimental results concerning normalized stress to eliminate the effects of density variations. A good correlation can also be seen from Fig. 6.29. The normalized stress σ_n was calculated as $\sigma_n = (\sigma_{eng}/\rho_r)^{3/2}$, where σ_{eng} is engineering stress and ρ_r is the relative density. The mechanical properties (e.g., densification strain and plateau stress) are reproducible at low and medium porosity, whereas a more significant

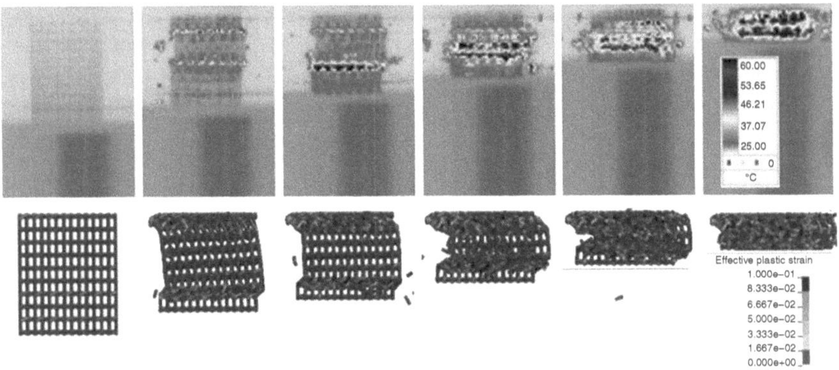

Fig. 6.28 ***Comparison of crushing behavior (short low-porosity specimens, loading direction X3, and strain increment: 20%): infrared images (*****top*****) and computational results (*****bottom*****).*** *(From N. Novak, M. Vesenjak, L. Krstulović-Opara, and Z. Ren. 2018. Mechanical characterisation of auxetic cellular structures built from inverted tetrapods. Composite Structures 196: 96–107).*

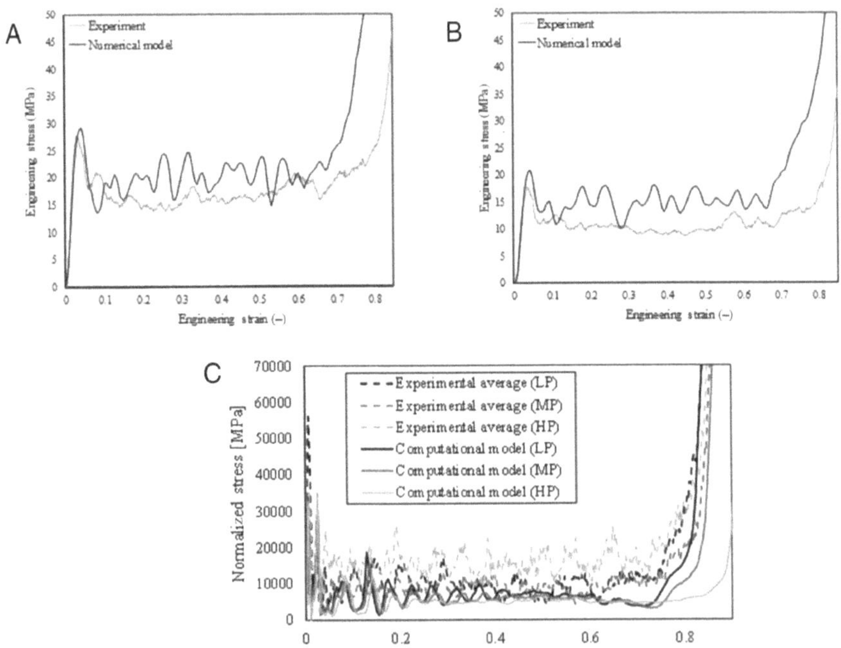

Fig. 6.29 ***Comparison of computational and experimental results for long specimens: low (A), middle porosity (B), and high strain rate (C).*** *(From N. Novak, M. Vesenjak, L. Krstulović-Opara, and Z. Ren. 2018. Mechanical characterisation of auxetic cellular structures built from inverted tetrapods.* Composite Structures *196: 96–107 and N. Novak, K. Hokamoto, M. Vesenjak, and Z. Ren. 2018. Mechanical behaviour of auxetic cellular structures built from inverted tetrapods at high strain rates.* International Journal of Impact Engineering *122: 83–90).*

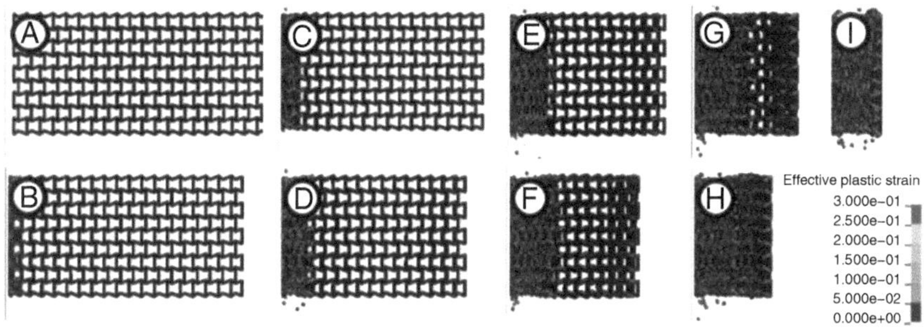

Fig. 6.30 ***A crushing sequence of long specimens of middle porosity (strain increment: 10%).*** *(From N. Novak, K. Hokamoto, M. Vesenjak, and Z. Ren. 2018. Mechanical behaviour of auxetic cellular structures built from inverted tetrapods at high strain rates.* International Journal of Impact Engineering *122: 83–90).*

deviation is observed at high porosity. The latter is attributed to increased influence of the manufacturing process due to imperfections at the strut surfaces influencing the deformation behavior of smaller struts, which has already been analyzed in detail by Novak et al. [17].

The computational model of the chiral auxetic cellular structure was validated with uniaxial compression testing. The experimental and computational deformation behavior of specimens is shown in Fig. 6.31, where an excellent agreement can be observed.

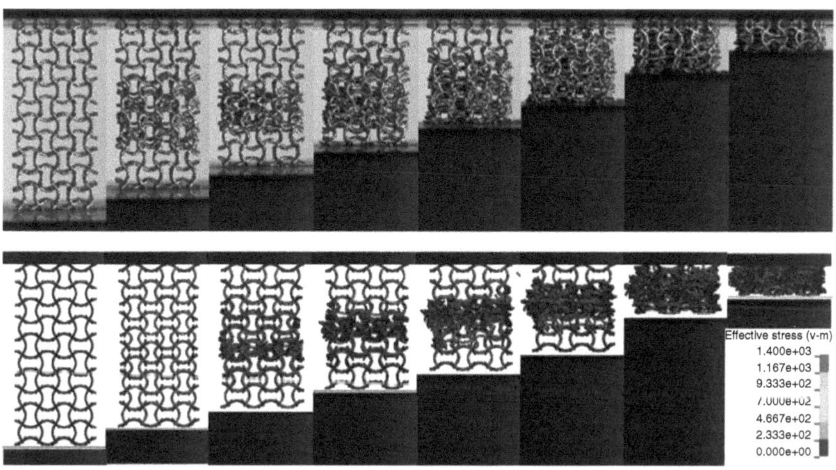

Fig. 6.31 ***Comparison of experimental (top) and computational results (bottom) (strain increment: 10%).*** *(From N. Novak, L. Starčevič, M. Vesenjak, and Z. Ren. 2019. Blast response study of the sandwich composite panels with 3D chiral auxetic core. Composite Structures 210: 167–178).*

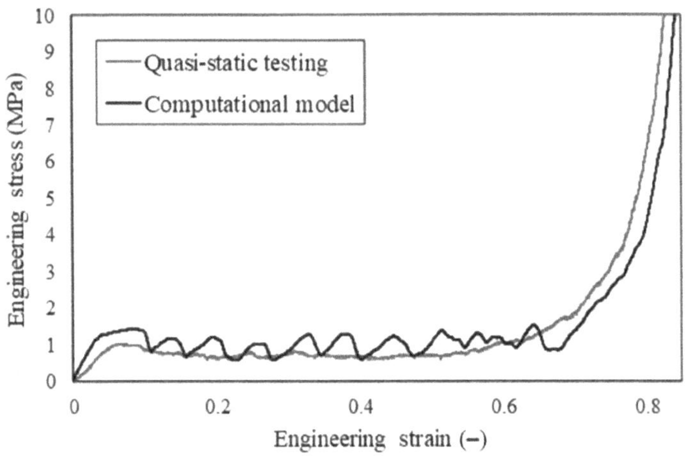

Fig. 6.32 ***Evolution of the reaction force on the support plate obtained from the experimental test and computational simulation.*** *(From N. Novak, L. Starčevič, M. Vesenjak, and Z. Ren. 2019. Blast response study of the sandwich composite panels with 3D chiral auxetic core.* Composite Structures *210: 167–178).*

The validation of the material model was performed by comparing the reaction force on the loading plate (Fig. 6.32), measured during the experimental test and computational simulation. A good agreement can be observed.

6.6 Conclusions

The mechanical behavior of various types of cellular structures from quasi-static to high strain rate loading was presented in this chapter. The techniques and approaches to quasistatic, medium strain rate, and high strain rate experimental testing are discussed in detail based on testing of four groups of cellular structures herein: UniPore cellular structures made from copper, open- and closed-cell aluminum foams, and auxetic structures (tetrapod and chiral made from titanium and copper). The experimental results of these structures under compression loading are compared in Fig. 6.33 with respect to the specific energy absorption (SEA) capacity at 50% strain. The darker shading colors in Fig. 6.33 correspond to lower specimen porosity in each analyzed group of cellular structures, whereas lighter shading denotes higher-porosity specimens. It can be observed that porosity and base material strongly influence the SEA capacity. Structures with low porosity can absorb more mechanical energy through deformation than structures with higher porosity that exhibit lower stiffness. The same conclusion can also be reached for base materials, where

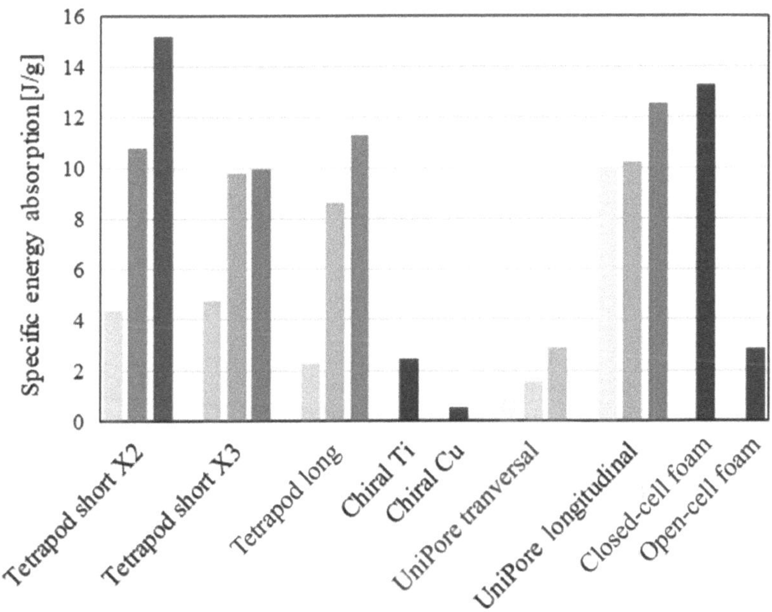

Fig. 6.33 ***Comparison of specific energy absorption capabilities of different cellular structures.***

materials with a higher Young modulus show higher SEA capacity. In absolute terms, the best consistent SEA at 50% strain can be observed for closed-cell foams and UniPore structures. It should be noted that in UniPore structures, densification already appears at 50% strain (see Fig. 6.5) and the shown SEA is thus the total SEA for this type of structure. Closed-cell foams (see Fig. 6.13) and other analyzed cellular structures densify at much higher strains, which contributes to higher total SEA capacity of these cellular structures.

The computational simulations are a great tool for better understanding of the deformation behavior of various cellular structures. Different approaches are presented in this chapter, from simpler and fast homogenized computational models to an advanced detailed and time-consuming approach, where a whole cellular structure is modeled with volume finite elements. A further improvement of computational models is imminent with the development of computational power and software capabilities, which will also enable inclusion of defects caused by the fabrication procedures and mesoscale modeling approaches, where the properties of the metal's crystallography will be taken into account as well. Consequently, it will be possible to study the deformation behavior and mechanical response of cellular structures even more precisely in future applications.

References

[1] L.J. Gibson, M.F. Ashby, Cellular Solids: Structure and Properties, Cambridge University Press, Cambridge, UK, 1997.
[2] M.F. Ashby, A. Evans, N.A. Fleck, L.J. Gibson, J.W. Hutchinson, H.N.G. Wadley, Metal Foams: A Design Guide, Elsevier Science, Burlington, MA, 2000.
[3] D. Lehmhus, M. Vesenjak, S. de Schampheleire, T. Fiedler, From stochastic foam to designed structure: balancing cost and performance of cellular metals., Materials (Basel) 10 (922) (2017) 1–32.
[4] J. Banhart, Manufacture, characterization and application of cellular metals and metal foams, Progress in Materials Science 46 (6) (2001) 559–632.
[5] I. Duarte, N. Peixinho, A. Andrade-Campos, R. Valente, Special issue on cellular materials, Science and Technology of Materials 30 (1) (2018) 1–3.
[6] D.S. Schwartz, S.D. Shih, A.G. Evans, H.N.G. Wadley, Porous and Cellular Materials for Structural Applications. Vol. 521, Materials Research Society, Warrendale, PA, 1998.
[7] I. Duarte, M. Vesenjak, L. Krstulović-Opara, Z. Ren, Static and dynamic axial crush performance of in-situ foam-filled tubes, Composite Structures 124 (2015) 128–139.
[8] S. Tanaka, K. Hokamoto, S. Irie, T. Okano, Z. Ren, M. Vesenjak, and S. Itoh, High-velocity impact experiment of aluminum foam sample using powder gun, Measurement 44 (10) (2011) 2185–2189.
[9] M. Vesenjak, A. Kovačič, M. Tane, M. Borovinšek, H. Nakajima, Z. Ren, Compressive properties of lotus-type porous iron., Computational Materials Science 65 (2012) 37–43.
[10] M. Vesenjak, K. Hokamoto, S. Matsumoto, Y. Marumo, Z. Ren, Uni-directional porous metal fabricated by rolling of copper sheet and explosive compaction, Materials Letters 170 (2016) 39–43.
[11] W. Sanders, L. Gibson, Mechanics of BCC and FCC hollow-sphere foams, Materials Science and Engineering: A 352 (1-2) (2003) 150–161.
[12] M. Vesenjak, T. Fiedler, Z. Ren, A. Öchsner, Dynamic behaviour of metallic hollow sphere structures, in: A. Öechsner, C. Augustin (Eds.), Multifunctional Metallic Hollow Sphere Structures, Engineering Materials Book Series. Springer, Berlin, Germany, 2009, pp. 137–158.
[13] K. Stöbener, J. Baumeister, G. Rausch, M. Rausch, Forming metal foams by simpler methods for cheaper solutions, Metal Powder Report 60 (1) (2005) 12–16.
[14] Y.C. Fung, Foundations of Solid Mechanics, Prentice Hall, Englewood Cliffs, NJ, 1965.
[15] K.E. Evans, M.A. Nkansah, I.J. Hutchinson, S.C. Rogers, Molecular network design, Nature 353 (1991) 124.
[16] N. Novak, M. Vesenjak, I. Duarte, S. Tanaka, K. Hokamoto, L. Krstulović-Opara, B. Guo, P. Chen, and Z. Ren, Compressive behaviour of closed-cell aluminium foam at different strain rates, Materials (Basel) 12 (24) (2019) 4108.
[17] N. Novak, M. Vesenjak, L. Krstulović-Opara, Z. Ren, Mechanical characterisation of auxetic cellular structures built from inverted tetrapods, Composite Structures 196 (2018) 96–107.
[18] M. Borovinšek, M. Vesenjak, and Z. Ren. Estimating the base material properties of sintered metallic hollow spheres by inverse engineering procedure. Mechanics of Materials 100(2016) 22–30.
[19] M. Vesenjak, C. Veyhl, T. Fiedler, Analysis of anisotropy and strain rate sensitivity of open-cell metal foam, Materials Science and Engineering: A 541 (2012) 105–109.
[20] N. Novak, M. Vesenjak, Z. Ren, Computational simulation and optimization of functionally graded auxetic structures made from inverted tetrapods, Physica Status Solidi B 254 (12) (2017) 1600753.

[21] M. Nishi, M. Oshita, M. Ulbin, M. Vesenjak, Z. Ren, K. Hokamoto, Computational analysis of the explosive compaction fabrication process of cylindrical uni-directional porous copper, Metals and Materials International 24 (5) (2018) 1143–1148.
[22] International Organization for Standardization. Mechanical Testing of Metals–Ductility Testing–Compression Test for Porous and Cellular Metals, ISO 13314:2011. International Organization for Standardization, Geneva, Switzerland, 2011.
[23] L. Krstulović-Opara, M. Surjak, M. Vesenjak, Z. Tonković, J. Kodvanj, Ž. Domazet., Comparison of infrared and 3D digital image correlation techniques applied for mechanical testing of materials, Infrared Physics and Techology 73 (2015) 166–174.
[24] L. Krstulovi, M. Vesenjak, I. Duarte, Z. Ren, Infrared thermography as a method for energy absorption evaluation of metal foams, Materials Today: Proceedings 3 (4) (2016) 1025–1030.
[25] N. Novak, K. Hokamoto, M. Vesenjak, Z. Ren, Mechanical behaviour of auxetic cellular structures built from inverted tetrapods at high strain rates, International Journal of Impact Engineering 122 (2018) 83–90.
[26] K. Hokamoto, M. Vesenjak, Z. Ren, Fabrication of cylindrical uni-directional porous metal with explosive compaction, Materials Letters 137 (2014) 323–327.
[27] M. Vesenjak, K. Hokamoto, M. Sakamoto, T. Nishi, L. Krstulović-Opara, Z. Ren, Mechanical and microstructural analysis of unidirectional porous (UniPore) copper, Materials and Design 90 (2016) 867–880.
[28] M. Nishi, S. Tanaka, M. Vesenjak, Z. Ren, K. Hokamoto, Experimental and computational analysis of the uni-directional porous (UniPore) copper mechanical response at high-velocity impact, International Journal of Impact Engineering 136 (2020) 103409.
[29] M. Borovinšek, M. Vesenjak, M. Jože, Z. Ren, Computational reconstruction of scanned aluminum foams for virtual testing, Journal of the Serbian Society for Computational Mechanics 2 (2) (2008) 16–28.
[30] I. Duarte, M. Vesenjak, L. Krstulović-Opara, Compressive behaviour of unconstrained and constrained integral-skin closed-cell aluminium foam, Composite Structures 154 (2016) 231–238.
[31] I. Duarte, L. Krstulović-Opara, M. Vesenjak, Characterization of aluminium alloy tubes filled with aluminium alloy integral-skin foam under axial compressive loads., Composite Structures 121 (2015) 154–162.
[32] I. Duarte, M. Vesenjak, L. Krstulović-Opara, Dynamic and quasi-static bending behaviour of thin-walled aluminium tubes filled with aluminium foam, Composite Structures 109 (1) (2014) 48–56.
[33] L. Peroni, M. Avalle, M. Peroni, The mechanical behaviour of aluminium foam structures in different loading conditions, International Journal of Impact Engineering 35 (7) (2008) 644–658.
[34] M. Ulbin, M. Vesenjak, M. Borovinšek, I. Duarte, Y. Higa, K. Shimojima, and Z. Ren, Detailed analysis of closed-cell aluminum alloy foam internal structure changes during compressive deformation, Advanced Engineering Materials 20 (8) (2018) 1–8.
[35] M.A. Islam, A. D. Brown, P. J. Hazell, M. A. Kader, J. P. Escobedo, M. Saadatfar, S. Xu, D. Ruan, and M. Turner, Mechanical response and dynamic deformation mechanisms of closed-cell aluminium alloy foams under dynamic loading, International Journal of Impact Engineering 114 (2018) 111–122.
[36] V.S. Deshpande, N.A. Fleck, High strain rate compressive behaviour of aluminium alloy foams, International Journal of Impact Engineering 24 (3) (2000) 277–298.
[37] Z. Zheng, J. Yu, C. Wang, S. Liao, Y. Liu, Dynamic crushing of cellular materials: a unified framework of plastic shock wave models, International Journal of Impact Engineering 53 (2013) 29–43.
[38] X. Liu, J. Zhang, Q. Fang, H. Wu, Y. Zhang, Response of closed-cell aluminum foams under static and impact loading: experimental and mesoscopic numerical analysis., International Journal of Impact Engineering 110 (2017) 382–394.

[39] Q. Fang, J. Zhang, Y. Zhang, J. Liu, Z. Gong, Mesoscopic investigation of closed-cell aluminum foams on energy absorption capability under impact, Composite Structures 124 (2015) 409–420.
[40] Q. Fang, J. Zhang, Y. Zhang, H. Wu, Z. Gong, A 3D mesoscopic model for the closed-cell metallic foams subjected to static and dynamic loadings, International Journal of Impact Engineering 82 (2015) 103–112.
[41] S. Wang, Y. Ding, C. Wang, Z. Zheng, J. Yu, Dynamic material parameters of closed-cell foams under high-velocity impact, International Journal of Impact Engineering 99 (2017) 111–121.
[42] Y. Sun, Q.M. Li, T. Lowe, S.A. McDonald, P.J. Withers, Investigation of strain-rate effect on the compressive behaviour of closed-cell aluminium foam by 3D image-based modelling, Materials and Design 89 (2016) 215–224.
[43] C.Q. Xi, Q.M. Li, Meso-scale mechanism of compaction shock propagation in cellular materials, International Journal of Impact Engineering 109 (2017) 321–334.
[44] A. Xu, T. Vodenitcharova, K. Kabir, E.A. Flores-Johnson, M. Hoffman, Finite element analysis of indentation of aluminium foam and sandwich panels with aluminium foam core, Materials Science and Engineering: A 599 (2014) 125–133.
[45] N. Novak, M. Vesenjak, Z. Ren, Crush behaviour of auxetic cellular structures, Science of Technology and Materials 30 (1) (2018) 4–7.
[46] P. Yang, S.S. Shams, A. Slay, B. Brokate, and R. Elhajjar. 2015. Evaluation of temperature effects on low velocity impact damage in composite sandwich panels with polymeric foam cores. Composite Structures 129: 213–223.
[47] C. Yang, S. Kyriakides, Multiaxial crushing of open-cell foams, International Journal of Solids and Structures 159 (2019) 239–256.
[48] S. Gaitanaros, S. Kyriakides, A.M. Kraynik, On the crushing response of random open-cell foams, International Journal of Solids and Structures 49 (19-20) (2012) 2733–2743.
[49] S. Gaitanaros, S. Kyriakides, On the effect of relative density on the crushing and energy absorption of open-cell foams under impact, International Journal of Impact Engineering (2015).
[50] Y. Sun, Q.M. Li, Dynamic compressive behaviour of cellular materials: a review of phenomenon, mechanism and modelling, International Journal of Impact Engineering 112 (2018) 74–115.
[51] L. Xiao, W. Song, C. Wang, H. Tang, Q. Fan, N. Liu, J. Wang, et al., Mechanical properties of open-cell rhombic dodecahedron titanium alloy lattice structure manufactured using electron beam melting under dynamic loading, International Journal of Impact Engineering 100 (2017) 75–89.
[52] P.J. Tan, S.R. Reid, J.J. Harrigan, Z. Zou, S. Li, Dynamic compressive strength properties of aluminium foams. Part I—experimental data and observations, Journal of Mechanics and Physics of Solids 53 (10) (2005) 2174–2205.
[53] P.J. Tan, S.R. Reid, J.J. Harrigan, Z. Zou, S. Li, Dynamic compressive strength properties of aluminium foams. Part II—'shock' theory and comparison with experimental data and numerical models, Journal of Mechanics and Physics of Solids 53 (10) (2005) 2206–2230.
[54] I. Duarte, M. Vesenjak, M.J. Vide, Automated continuous production line of parts made of metallic foams, Metals (Basel) 9 (5) (2019) 531.
[55] T. Hipke, G. Lange, R. Poss, Taschenbuch für Aluminiumschäume, Alu Media, Düsseldorf, Germany, 2007.
[56] I. Duarte, M. Vesenjak, L. Krstulović-Opara, Variation of quasi-static and dynamic compressive properties in a single aluminium foam block, Materials Science and Engineering: A 616 (2014) 171–182.
[57] Y.T. Yao, M. Uzun, I. Patel, Workings of auxetic nano-materials, Journal of Achievements in Materials and Manufacturing Engineering 49 (2) (2011) 585–593.

[58] J. Schwerdtfeger, P. Heinl, R. F. Singer, and C. Körner. Auxetic cellular structures through selective electron beam melting. Physica Status Solidi B 247(2)(2010) 269–272.
[59] A. Yeganeh-Haeri, D.J. Weidner, J.B. Parise, Elasticity of α-cristobalite: a silicon dioxide with a negative Poisson's ratio, Science 257 (5070) (1992) 650–652.
[60] C. Lees, J.F. Vincent, J.E. Hillerton, Poisson's ratio in skin, Bio-Medical Materials and Engineering 1 (1) (1991) 19–23.
[61] W. Voigt, Lehrbuch der Kristallphysik (mit Ausschluss der Kristalloptik) (1928).
[62] K. Muto, R.W. Bailey, K.J. Mitchell, pecial requirements for the design of nuclear power stations to withstand earthquakes, Proceedings of the Institution of Mechanical Engineers 177 (1) (1963) 155–203.
[63] R.F. Almgren, An isotropic three-dimensional structure with Poisson's ratio = −1, Journal of Elasticity 15 (1985) 427–430.
[64] L.J. Gibson, M.F. Ashby, The mechanics of 2D cellular material, Proceedings of the Royal Society A 382 (1782) (1982) 25–45.
[65] R.S. Lakes, Foam structures with a negative Poisson's ratio, Science 235 (1987) 1038–1040.
[66] G.W. Milton, Composite materials with Poisson's ratios close to −1, Journal of Mechanics and Physics of Solids 40 (5) (1992) 1105–1137.
[67] J.N. Grima, K.E. Evans, Auxetic behavior from rotating squares, Journal of Materials Science Letters 19 (17) (2000) 1563–1565.
[68] T. Bückmann, N. Stenger, M. Kadic, J. Kascheke, A. Frölich, T. Kennerknecht, C. Eberl, M. Thiel, M. Wegener, Tailored 3D mechanical metamaterials made by dip-in direct-laser-writing optical lithography, Advanced Materials (Deerfield Beach, Fla.) 24 (20) (2012) 2710–2714.
[69] T.J. Horn, O.L.A. Harrysson, D.J. Marcellin-Little, H.A. West, B.D.X. Lascelles, R. Aman, Flexural properties of Ti6Al4V rhombic dodecahedron open cellular structures fabricated with electron beam melting, Additive Manufacturing 1-4 (2014) 2–11.
[70] E.A. Friis, R.S. Lakes, J.B. Park, Negative Poisson's ratio polymeric and metallic foams, Journal of Materials Science 23 (12) (1988) 4406–4414.
[71] S. Mohsenizadeh, R. Alipour, M. Shokri Rad, A.F Nejad, Z. Ahmad, Crashworthiness assessment of auxetic foam-filled tube under quasi-static axial loading, Materials and Design 88 (2015) 258–268.
[72] S. Hou, T. Liu, Z. Zhang, X. Han, Q. Li, How does negative Poisson's ratio of foam filler affect crashworthiness?, Materials and Design 82 (2015) 247–259.
[73] L. Yang, O. Harrysson, H. West, D. Cormier, Compressive properties of Ti–6Al–4V auxetic mesh structures made by electron beam melting, Acta Materialia 60 (8) (2012) 3370–3379.
[74] L. Yang, O. Harrysson, H. West, D. Cormier, Mechanical properties of 3D re-entrant honeycomb auxetic structures realized via additive manufacturing, International Journal of Solids and Structures 69-70 (2015) 475–490.
[75] J. Schwerdtfeger, F. Schury, M. Stingl, F. Wein, R.F. Singer, C. Körner, Mechanical characterisation of a periodic auxetic structure produced by SEBM, Physica Status Solidi B 249 (7) (2012) 1347–1352.
[76] O. Duncan, L. Foster, T. Senior, A. Alderson, T. Allen, Quasi-static characterization and impact testing of auxetic foam for sports safety applications, Smart Materials and Structures 25 (5) (2016) 054014.
[77] T. Fíla, et al., Impact testing of polymer-filled auxetics using split Hopkinson pressure bar, Advanced Engineering Materials 19 (10) (2017) 1700076.
[78] L. Jiang, H. Hu, Low-velocity impact response of multilayer orthogonal structural composite with auxetic effect, Composite Structures 169 (2016) 62–68.
[79] C. Qi, A. Remennikov, L.-Z. Pei, S. Yang, Z.-H. Yu, T.D. Ngo, Impact and close-in blast response of auxetic honeycomb-cored sandwich panels: experimental tests and numerical simulations, Composite Structures 180 (2017) 161–178.
[80] T. Strek, H. Jopek, M. Nienartowicz, Dynamic response of sandwich panels with auxetic cores, Physica Status Solidi B 252 (7) (2015) 1540–1550.

[81] L. Boldrin, S. Hummel, F. Scarpa, D. Di Maio, C. Lira, M. Ruzzene, C. D. L. Remillat, T. C. Lim, R. Rajasekaran, and S. Patsias, Dynamic behaviour of auxetic gradient composite hexagonal honeycombs, Composite Structures 149 (2016) 114–124.
[82] G. Imbalzano, P. Tran, T.D. Ngo, P.V.S. Lee, A numerical study of auxetic composite panels under blast loadings, Composite Structures 135 (2015) 339–352.
[83] S. Yang, C. Qi, D. Wang, R. Gao, H. Hu, J. Shu, A comparative study of ballistic resistance of sandwich panels with aluminum foam and auxetic honeycomb cores, Advances in Mechanical Engineering 5 (2013) 589216.
[84] C. Körner, Y. Liebold-Ribeiro, A systematic approach to identify cellular auxetic materials, Smart Materials and Structures 24 (2014) 025013.
[85] F. Warmuth, C. Körner, Phononic band gaps in 2D quadratic and 3D cubic cellular structures, Materials (Basel) 8 (12) (2015) 8327–8337.
[86] M. Wormser, F. Warmuth, C. Körner, Evolution of full phononic band gaps in periodic cellular structures, Applied Physics A 123 (2017) 661.
[87] F. Warmuth, F. Osmanlic, L. Adler, Fabrication and characterization of a fully auxetic 3D lattice structure via selective electron beam melting, Smart Materials and Structures 26 (8) (2017) 025013.

Index

Page numbers followed by '*t*' indicate tables and *f* indicate figures.

Q

R

S

T

U

V

W

FSC
www.fsc.org
MIX
Papier | Fördert gute Waldnutzung
FSC® C083411

Kolibri 360 GmbH
Ferdinand-Jühlke-Straße 7
99095 Erfurt, Deutschland
produktsicherheit@kolibri360.de